Spectral Domain Method for
Microwave Integrated Circuits

ELECTRONIC & ELECTRICAL ENGINEERING RESEARCH STUDIES

COMPUTER METHODS IN ELECTROMAGNETICS SERIES

Series Editor: **Professor J. B. Davies**
University College, London, England

1. Spectral Domain Method for Microwave Integrated Circuits*
   D. Mirshekar-Syahkal

*The computer programs listed in this book are available on 5¼" and 3½" disks, and on ¼" magnetic cartridge, obtainable from the author.

Spectral Domain Method for Microwave Integrated Circuits

D. Mirshekar-Syahkal
Department of Electronic Systems Engineering
University of Essex
Wivenhoe Park, Colchester, UK CO4 3SQ

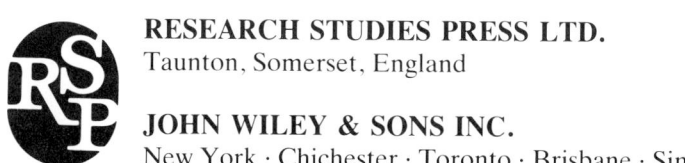

RESEARCH STUDIES PRESS LTD.
Taunton, Somerset, England

JOHN WILEY & SONS INC.
New York · Chichester · Toronto · Brisbane · Singapore

RESEARCH STUDIES PRESS LTD.
24 Belvedere Road, Taunton, Somerset, England TA1 1HD

Copyright © 1990, by Research Studies Press Ltd.

All rights reserved.

No part of this book may be reproduced by any means,
nor transmitted, nor translated into a machine language
without the written permission of the publisher.

Marketing and Distribution:

Australia and New Zealand:
JACARANDA WILEY LTD.
GPO Box 859, Brisbane, Queensland 4001, Australia

Canada:
JOHN WILEY & SONS CANADA LIMITED
22 Worcester Road, Rexdale, Ontario, Canada

Europe, Africa, Middle East and Japan:
JOHN WILEY & SONS LIMITED
Baffins Lane, Chichester, West Sussex, England

North and South America:
JOHN WILEY & SONS INC.
605 Third Avenue, New York, NY 10158, USA

South East Asia:
JOHN WILEY & SONS (SEA) PTE LTD.
37 Jalan Pemimpin #05-04
Block B Union Industrial Building, Singapore 2057

Library of Congress Cataloging in Publication Data

Mirshekar-Syahkal, D. (Dariush), 1951–
    Spectral domain method for microwave integrated circuits / D. Mirshekar-Syahkal.
        p.   cm. — (Electronic & electrical engineering research studies. Computer methods in electromagnetics series ; 1)
    Includes bibliographical references.
    ISBN 0-471-92684-1 (Wiley)
    1. Microwave integrated circuits—Mathematics.   I. Title.
II. Series.
TK7876.M592   1990
621.381'32—dc20   89-21499
                                                               CIP

British Library Cataloguing in Publication Data

Mirshekar – Syahkal, D.
    Spectral domain method for microwave integrated circuits.
    1. Electronic equipment integrated circuits, microwave.
    Domain analysis spectral
    I. Title   II. Series
    621. 3815

    ISBN 0 86380 099 8

    ISBN 0 86380 099 8 (Research Studies Press Ltd.)
    ISBN 0 471 92684 1 (John Wiley & Sons Inc.)

Printed in Great Britain by SRP Ltd., Exeter

*To my parents,
my wife Tabassom,
and our children Negar and Bahar*

CONTENTS

Foreword		**xi**
Preface		**xiii**
Chapter 1	**INTRODUCTION**	**1**
1.1	Microwave Integrated Circuits (MICs)	1
1.2	Electromagnetic Field Problems in MICs	4
1.3	The Spectral Domain Technique	7
	References	9
Chapter 2	**BASIC EQUATIONS AND CONCEPTS**	**11**
2.1	Maxwell's Equations	11
2.2	Constitutive Relations	12
2.3	Boundary Conditions	12
2.4	Edge Condition	14
2.5	Potential Functions	19
	References	22
Chapter 3	**INTRODUCTION TO THE SPECTRAL DOMAIN TECHNIQUE**	**23**
3.1	Microstrip Line and the Spectral Domain Technique	23

3.2	Spectral Domain Formulation of Microstrip Lines	25
3.2.1	Potential Functions in the Fourier Domain	26
3.2.2	Boundary Conditions in the Space and in the Fourier Domain	29
3.2.3	Final Equations in the Fourier Domain	31
3.2.4	Final Equations in the Space Domain	34
3.3	Methods of Solution of the Final Equations	38
3.3.1	Point Matching (Point Collocation) Method	38
3.3.2	The Method of Moments	43
3.3.2.1	Galerkin Technique in the Space Domain	45
3.3.2.2	Galerkin Technique in the Spectral Domain	47
	References	49

Chapter 4 EFFICIENT COMPUTING AND APPROXIMATIONS IN THE SPECTRAL DOMAIN TECHNIQUE   53

4.1	Basis Functions	53
4.2	Approximate Solutions	60
4.3	Relations among Elements of the Coefficient Matrix	63
4.4	Alternative Spectral Domain Green's Functions	66
4.5	Acceleration of the Computation of the Matrix Elements	70
	References	72

Chapter 5 QUASI-TEM ANALYSIS BY THE SPECTRAL DOMAIN TECHNIQUE AND SOLUTION OF MULTISTRIP TRANSMISSION LINES   75

5.1	Quasi-TEM Spectral Domain Formulation of the Microstrip Line	75
5.2	Spectral Domain Solution of Multistrip Single-Substrate Transmission Lines	80
5.2.1	Asymmetric Coupled Microstrip Lines	81
5.2.2	Multiconductor Microstrip Lines	84
	References	87

Chapter 6	SPECTRAL DOMAIN SOLUTION OF MULTILAYER MULTICONDUCTOR PLANAR TRANSMISSION LINES	89
6.1	Spectral Domain Solution of Multilayer Transmission Lines with Coplanar Conductors Using Transfer Matrix Approach	91
6.2	Three-Layer Planar Transmission Lines with Coplanar Conductors	100
6.2.1	Coplanar Waveguides	104
6.2.2	Inverted Microstrip Lines	107
6.2.3	Finlines	108
6.3	Spectral Domain Solution of Multilayer Transmission Lines with Multilayer Conductors	117
6.3.1	Coupled Strip-Finline Structure	121
6.4	Other Approaches for Generating the Spectral Domain Green's Functions of Multilayer Multiconductor Planar Transmission Lines	129
	References	137
Chapter 7	MISCELLANEOUS APPLICATIONS OF THE SPECTRAL DOMAIN TECHNIQUE	141
7.1	Solution of Microstrip-Type Resonators	141
7.2	Solution of Microstrip Patch Antennas	150
7.3	Solution of Planar Structures with Periodic Metallisation	160
7.4	Solution of Scattering from Planar Structures	168
7.5	Solution of Planar Structures with Lossy and/or Anisotropic Substrates	172
	References	179
Appendix I	FOURIER TRANSFORMS	185
I.1	Expansion of Periodic Functions in terms of Trigonometric Functions (Fourier Series)	185
I.2	Fourier Series of Odd and Even Functions	186

I.3	Half-Range Expansions	187
I.4	Finite Fourier Transform	188
I.5	Fourier Integral and Fourier Transform	189
I.6	Fourier Transforms of the Derivatives of a Function	189
I.7	Parseval's Identities	190
	References	190

Appendix II LINEAR INTEGRAL EQUATIONS   191

II.1	Classification	191
II.2	Types of Kernels	192
II.3	Green's Function and its Role in Integral Equations	194
	References	196

Appendix III COMPUTER PROGRAMS   199

III.1	Computer Program **MRST1**	200
III.2	Subroutine **MGNTW**	210
III.3	Subroutine **ANPDL**	210
III.4	Subroutine **MSCL2**	211
III.5	Listing of **MRST1**	212
III.6	Sample Input and Output	224
III.7	Listing of **MGNTW**	228
III.8	Listing of **ANPDL**	231
III.9	Listing of **MSCL2**	234
	References	237

Index   239

Foreword

Planar transmission lines provide the chief way of guiding microwaves (and indeed lower frequencies) within components and enclosed systems. Since the introduction of stripline and microstrip in the early 1950's, the use of planar forms of waveguide has seen a steady evolution, expansion and improvement. In the pursuit of better performance (in terms of compactness, losses, reproducibility) and more flexible structures (to allow for wide variety of components and convenient interconnections) the range of basic guides has grown. From microstrip has grown coplanar waveguide, inverted microstrip, suspended microstrip, slotline (its dual form) and finline, - and very many variants. They all offer their different advantages, and monolithic microwave integrated circuits (MMICs) have benefited from this rich variety of structures.

A key part of this progress has been in the availability of electromagnetic theory to provide adequate design capability, so that the engineer's needs can be translated into a mask and resulting component in (ideally) a one-step design. Among the various theoretical methods available, ones that take advantage of the special geometry are preferred for their efficiency, and the spectral domain method has emerged as the most commonly used. It has the scope of providing anything from quick and approximate results to the most accurate (though more computer-intensive) results, where the "user" can choose their trade-off between accuracy and computing time.

Accounts of the theory have appeared in the journals, over the years, as the various steps have evolved and improved. But without exception, the accounts leave out most of the detail and "rationale".

Dr. Mirshekar's text is aimed at providing everything from the basic theory to the latest "state-of-the-art", for this central method of analysing planar waveguides and structures.

He also provides here complete computer programs for a variety of microstrip, finline, coplanar waveguide, and slot-lines, including some in "symmetric coupler" form. In Fortran source form, these are equally suitable for immediate running on modest personal computers or incorporation in C.A.D. packages.

Prof. J. B. Davies
March 1989

Preface

The spectacular growth of Microwave Integrated Circuit (MIC) technology in the past decade has paved the way for further miniaturisation of microwave systems and subsystems. Articles recently published by investigators are indicative of significant achievements in the production of low-cost microwave chips.

In parallel with this rapid development, the complexity of MIC structures has posed a new challenge to electromagnetic field theoreticians. Various mathematical techniques have been applied to find accurate but easy solutions to MIC problems. Due to the inherent complex nature of the electromagnetic field in MIC structures, almost invariably all the mathematical techniques capable of rigorously predicting the basic circuit parameters are either numerical or pseudo-numerical techniques. Therefore, the quality of these techniques can be assessed by means of a set of criteria such as computing efficiency, ease of use, generality and reliability.

The Spectral Domain Technique (SDT) which is widely used for the analysis of planar microwave integrated circuits, has proved so far to possess qualities of a good pseudo-numerical method. Perhaps the first planar structures benefiting from the spectral domain treatment were microstrip and slot lines. A quick survey of the literature shows that the range of applications of this pseudo-numerical technique has been significantly extended, covering analysis of planar couplers, filters, antenna arrays and even GaAs planar structures.

From the mathematical point of view, the spectral domain technique is simply an integral transformation method. Under special circumstances the differential equations governing the behaviour of a system can be transformed into a new domain: the spectral domain, where the transformed equations have an easy solution. The actual solution is then obtained using an inverse transformation. Normally in boundary value problems the above process results in one or more integral equations in the original domain, whose solution can be sought by a standard numerical technique. The numerical efficiency of the solution, however, depends on the choice of the technique for converting the integral equations into the final set of linear equations.

Since the first publication on the spectral domain technique, many papers have appeared in microwave journals and conference proceedings explaining new applications or extensions of the technique. But it is interesting to note that none of these papers give a clear and detailed account of the subject. Therefore new users with little mathematical background find themselves confronted with all ranges of difficulties, hindering them from an effective application of the spectral domain technique to their problems. It has also been learnt that review papers presented on the subject are not of considerable help, and in some cases they even cause confusion.

Against this background it has been felt for some time that a coherent, detailed and easily digestible treatment of the spectral domain technique applicable to microwave planar structures is required. This book is written to serve this purpose. An attempt has been made to keep the book self-contained. Therefore, the necessary background mathematics and electromagnetic theory are included in the book. However, because the book is aimed at bridging the gap between the theory and practice, mathematical concepts are only elaborated here, leaving proofs of mathematical theorems to appropriate references. Examples presented in the book are deliberately selected from well-known MIC planar structures.

The book starts with an introductory chapter consisting of a brief review of microwave integrated circuits, planar structures and their associated field problems. Also in this chapter the role of the spectral domain technique in the analysis and design of many planar structures is mentioned. In Chapter 2, Maxwell's equations and boundary conditions are presented, followed by a discussion on the edge conditions including examples and illustrations. Chapter 3 introduces the spectral domain technique through the rigorous hybrid-mode formulation of the microstrip line. Not only does this way of presenting the spectral domain technique immediately reveal an application of the technique, but it is historically consistent with its development. In Chapter 3, the mathematical principles used in the spectral domain technique are gradually presented. Chapter 4 discusses various points on the computational side of the technique in general. In this chapter particular emphasis is placed upon the enhancement of the computational efficiency. Approximate solutions are also studied in Chapter 4. Chapter 5 consists of two main parts. The first part presents the quasi-TEM spectral domain technique. As in Chaper 3, this is accomplished by applying the technique to the microstrip line. Consequently, a direct comparison of the quasi-TEM formulation and the hybrid formulation of the microstrip line becomes possible. The second part of Chapter 5 is devoted to the solution of multiconductor single substrate transmission lines. In this part, direct applications of some of the equations derived in Chapters 3, 4 and the first part of Chapter 5 can be seen. Chapter 6 contains the more formal and general treatment of the spectral domain technique. In this chapter, planar transmission lines are assumed to have multilayer multiconductor structures. Detailed accounts of the "transfer matrix approach" and the "immittance approach" in the spectral domain technique are presented in Chapter 6. This chapter also embodies many useful mathematical expressions which can be directly used in the analysis of various planar structures. In this connection, several examples are presented. Chapter 7 presents miscellaneous applications of the spectral domain technique. It includes outline analyses of planar resonators, discontinuities, radiators, periodic structures and planar structures containing anisotropic or lossy materials. The importance of this chapter lies in the fact that the selected examples are representative of a broad class of microwave integrated circuit components. There are three appendices in the book. Appendix I is devoted to the Fourier expansions and transforms. Appendix II briefly discusses linear integral equations. Appendix III contains several useful computer codes in Fortran which are based on the spectral domain technique. These programs can be used to analyse many MIC transmission lines including various microstrip lines, coplanar waveguides, various finlines etc. These programs are available on 1/4" magnetic cartridge, 5 1/4" standard floppy disk and 3 1/2" microdisk and can be supplied by the author on request.

I would like to express my sincere thanks to Prof. J. Brian Davies, the series editor, for his meticulous editing and constructive comments throughout the preparation of the book. My thanks are also due to Mrs. L. Burnand for her expert and cheerful typing of the manuscript. To the companies that provided the photographs I am most appreciative. Also, I wish to acknowledge the help of Mr. H. Hassani, Mr. S. H. H. Sadeghi and Mr. M. L. Sansom in the preparation of some of the illustrations. Finally, I must specially thank my wife, Dr. Tabassom Malek Mirshekar-Syahkal who revised manuscripts and proof-read the entire book.

D. Mirshekar-Syahkal
June 1989

Chapter 1

INTRODUCTION

The past decade has witnessed a rapid development of microwave integrated circuits (MICs). Design and development of these circuits pose various electromagnetic field problems. Many of these problems can be analysed using the spectral domain technique. The role of this technique in the computer aided design (CAD) of MICs has been very significant. This powerful technique is now central to some CAD packages available commercially or developed as in-house tools for the design of MICs.

1.1 Microwave Integrated Circuits (MICs)

Microwave integrated circuits can be viewed as the extension of printed circuits at high frequencies. Present types of MICs have gradually evolved as the demand for very compact microwave systems and subsystems grew. In hybrid forms, almost all passive elements of an MIC are developed on a single or multilayer substrate. Active components such as transistors, diodes etc. are soldered or bonded to the circuit at their appropriate positions. The construction of microwave integrated circuits is usually planar. In MIC technology, the term 'planar structure' applies to flat stratified structures. In these structures, the circuit metallisations appear at the interfaces of layers of other materials involved in the structure, Fig. 1.1.

In many MICs, substrates and superstrates are usually lowloss dielectrics. In a more advanced type of MICs, monolithic microwave integrated circuits (MMICs), semi-insulating substrates such as GaAs are employed. Such a substrate is a suitable environment for the fabrication of both active and passive circuit components [1]. A few examples of MICs and MMICs are shown in Figs. 1.2-4.

CHAP. 1 INTRODUCTION

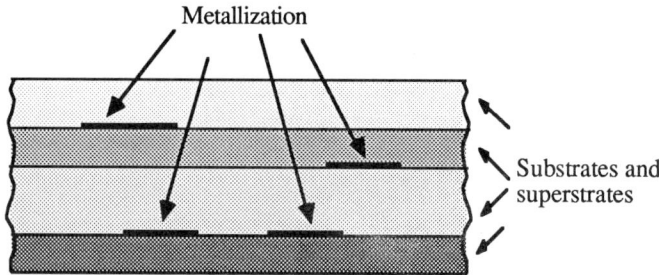

Fig. 1.1 Cross-section of a planar structure. Substrates and superstrates may be dielectrics, ferrites etc. depending on the circuit requirements.

In Fig. 1.2, a mixer developed on a Duroid substrate is shown. Duroid is a soft substrate which, like many other soft substrates [2], exhibits some degree of anisotropy. Although the anisotropic behaviour of soft substrates could lead to some complexities in the MIC design, they are widely used for their low cost and flexible mechanical properties. In the circuit shown in Fig. 1.2, the microstrip line serves as interconnection media, transformer and filter elements. In this circuit active elements are bonded to the microstrip lines. It is clear that the circuit resembles low-frequency printed circuits.

In Fig. 1.3, two microwave integrated circuits (Q-band mixers) based on the finline technology, are shown. In these circuits the substrates are clamped between the two sides of the standard waveguide WG-22. Finline technology was introduced in early 1970 and it is now an established technique for the development of MICs at frequencies above 20 GHz. At these frequencies, for various reasons such as high losses, compatibility problems with waveguides, intolerable line-widths etc., microstrip lines are less favoured and finlines are used instead [3]. However, fabrication processes for the development of circuits employing finlines or microstrip lines are the same.

In Fig. 1.4, a monolithic microwave integrated circuit (MMIC) is shown. The circuit which is a distributed amplifier employs GaAs substrate. Interconnections in these circuits are microstrip lines. In other forms, MMICs may employ coplanar

CHAP. 1 INTRODUCTION 3

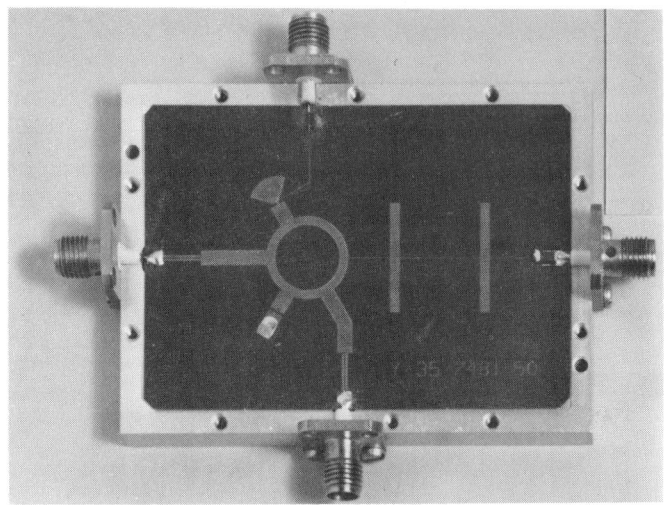

Fig. 1.2 A 12 GHz microstrip mixer. (Courtesy of GEC-Marconi Research Centre, Great Baddow, Essex, UK.)

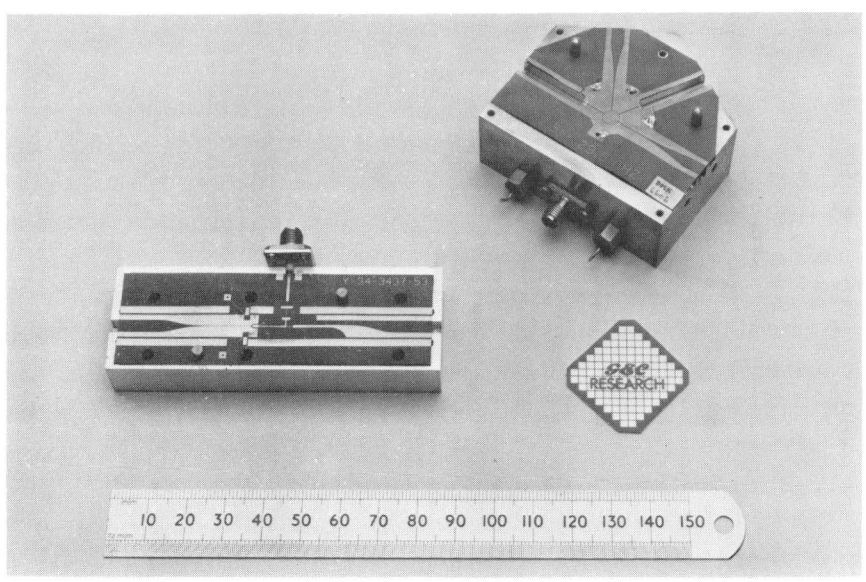

Fig. 1.3 Two Q-band finline mixers. (Courtesy of GEC-Marconi Research Centre, Great Baddow, Essex, UK.)

CHAP. 1 INTRODUCTION 4

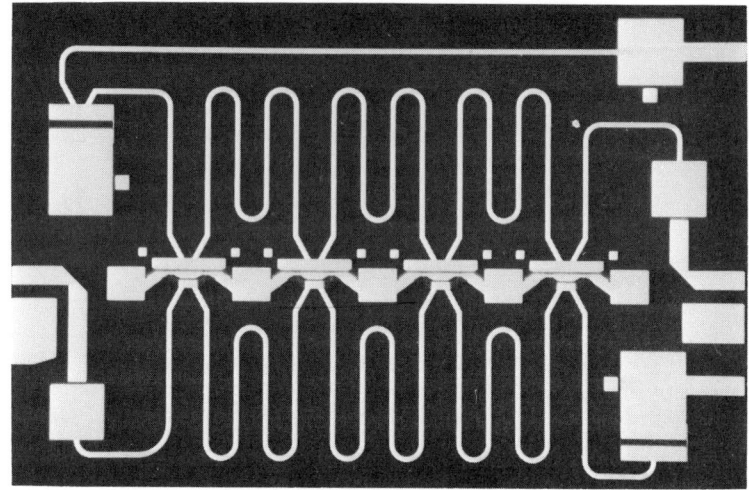

Fig. 1.4 A 2-6 GHz MMIC distributed amplifier on GaAs. (Courtesy of Philips Research Laboratories, Redhill, Surrey, UK.)

waveguides in place of microstrip lines. In those cases, the ground plane is easily accessible because it is etched on the same face of the substrate along with other required metallisations. Unlike the two previous MICs, the active elements of the circuit shown in Fig. 1.4 are made within the GaAs substrate using some deposition process [1].

It is worth emphasizing again that the three examples shown in Figs. 1.2-4 are all planar structures. However, note that with the recent development of MICs on cylindrical structures [4-6], the definition of planar structures conveys a broader concept now.

1.2 Electromagnetic Field Problems in MICs

In this book, the intention is not to examine circuit aspects of MICs, but to concentrate on a mathematical method, **the spectral domain technique**, which has been evolved in connection with the development of MICs. The chief purpose of the book is to introduce the reader to this technique and its application areas. Since its introduction in the analysis of microstrip lines [7], the spectral domain technique has been used in many occasions to tackle electromagnetic field problems associated with planar structures. Accurate design of many planar components has been due to availability of this method.

Let us now examine some of the electromagnetic field problems associated with MICs. Let us consider a simple structure first, the microstrip line. An accurate design of this line demands accessibility to design data in which the frequency effect as well as the effects of the physical and electrical parameters of the line are accurately reflected. Determination of such accurate design data is by no means a trivial task. Numerous papers published on this subject are indicative of the difficulties. The main problem in an accurate analysis of the microstrip line and indeed in the analysis of planar transmission lines lies in the fact that these structures involve mixed boundaries. Therefore, not only must all the six components of the field be taken into account in the formulation of a planar transmission line, but due to the presence of the strips at some interfaces, the final equations must be solved numerically, since they are rather complicated and the solution cannot be given in closed-form. The complexity of the solution increases if the substrates/superstrates are anisotropic materials.

When a discontinuity occurs along a planar transmission line or when the transmission line has to bend in order to connect the two parts of a circuit (see Figs. 1.2-4), some of the energy travelling along the line is reflected. Optimisation of the performance of a microwave integrated circuit calls for the initial characterisation of discontinuities, junctions, bends etc. existing in the circuit. A characterised discontinuity, junction, bend etc. is often represented by an approximate lump circuit model (a circuit including capacitors and inductors), Fig. 1.5. It is well-known that this characterisation is usually involved, particularly when the structures are in planar forms.

Radiation is also another electromagnetic problem associated with a category of planar structures. Open planar structures involving discontinuities radiate. For planar radiators such as microstrip and slot antennas, dimensions and materials should be selected so that the radiation is maximized. An accurate assessment of radiation from a planar structure depends on the accuracy of the solution of the associated electromagnetic problem. When dealing with planar radiators, a further electromagnetic problem to be addressed is the determination of the input impedance. Also problems of coupling among planar radiators when they are in an array

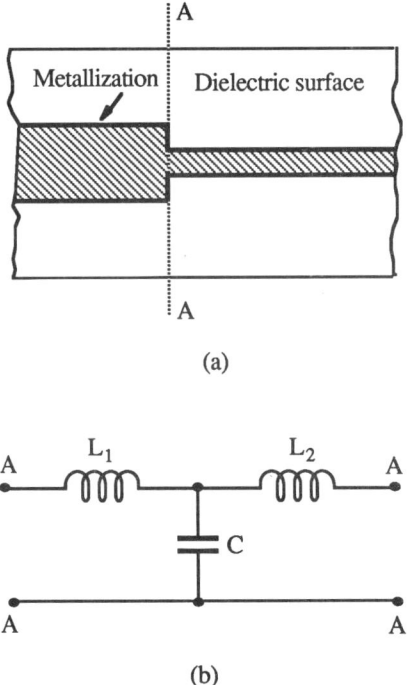

Fig. 1.5 A discontinuity in the microstrip line; (a) plan view and (b) its equivalent circuit.

arrangement have been under investigation for some time.

The electromagnetic problems alluded above are just a small selection of problems (although the fundamental ones) associated with MICs. In advanced computer aided design (CAD) packages developed for MICs, dimensions of the circuit components are generated from data-bases containing the results of accurate analyses of electromagnetic field problems associated with MICs. There are also CAD packages which employ approximate closed-form expressions for the same purpose. In principle, the former CAD packages are advantageous, since their data-bases can be updated for a new design range with little effort. However, both categories of the packages usually offer restricted design ranges. This is mainly due to the computer processing power limitation, since the actual analysing programs cannot be run in the middle of a CAD process (otherwise the process cannot be completed in a reasonable time). Furthermore, a program capable of the analysis of every perceivable MIC

CHAP. 1 INTRODUCTION 7

component is not feasible. Therefore, for the design of a planar component with a new shape, the designer should be able to develop the initial analysis and the associated computer program.

1.3 The Spectral Domain Technique

The spectral domain technique whose principles and applications are detailed in this book has proved itself as an elegant tool for the analysis of many electromagnetic problems associated with MICs. This technique is simple, powerful and computationally efficient when properly implemented. There are now CAD packages for the design of many MICs components which employ this technique as their data-base generator [8-9]. It has also been used in order to develop approximate closed-form expressions for the analysis and design of some planar structures [10-11].

One of the outstanding features of the spectral domain technique is its simplicity. A microwave engineer with a mathematical knowledge at the graduate level should be able to master the subject in a short time.

As the name 'Spectral Domain Technique' suggests, this technique involves the application of the Fourier transform. Under certain conditions, partial differential equations governing the behaviour of planar structures can be transformed into the Fourier domain. In this domain, the resulting differential equations have easy solutions. The philosophy behind the technique dates back to J.B. Fourier himself, although the advent of powerful computers have now made the full utilisation of the Fourier analysis feasible in different scientific branches. In the context of planar structures, the spectral domain technique was first introduced by Itoh et al [7] in 1973 for the analysis of the microstrip line. In the following years until 1976, new but basic applications of the spectral domain technique were reported by Jansen [12-13], Rahmat-Samii et al [14], Itoh et al [15] and many other investigators. These applications include analysis of various planar resonators, coupled microstrip lines, microstrip discontinuities etc.

In 1977, a generalised form of the spectral domain technique for applications to multilayer multiconductor planar transmission lines was proposed by Davies et al [16]. The technique employs a transfer matrix method in order to account for the extra layers of dielectrics in the analysis. Later in 1980, an alternative approach, the spectral domain immittance approach, for the spectral domain analysis of multilayer multiconductor transmission lines was introduced by Itoh [17]. Since then, various refinements have been made to the spectral domain technique in order to increase its numerical efficiency and its capabilities in dealing with more difficult MIC problems. The technique can now be reliably used to analyse planar radiators, periodic structures, finline structures, planar structures on semi-insulating and semi-conductor substrates etc. In connection with planar transmission lines, many of the established papers can conveniently be found in the book of selected reprints [18]. Further applications of the spectral domain technique in the analysis and design of new MIC components are still emerging. This can be clearly seen by examining recent issues of advanced journals in the field.

Although the spectral domain technique has many attractive features, one shortcoming of the technique is its inability to deal with planar structures employing thick conductors, Fig. 1.6. Therefore, at frequencies where the wavelengths become comparable with the metallisation thickness, the results from the spectral domain technique must be treated cautiously.

Before presenting a detailed account of the spectral domain technique and its applications, a brief review of the basic principles and concepts in the electromagnetic field theory is given in the following chapter. These are required for the later developments presented in the book. The chapter can, however, be skipped if the reader is already acquainted with these concepts and principles.

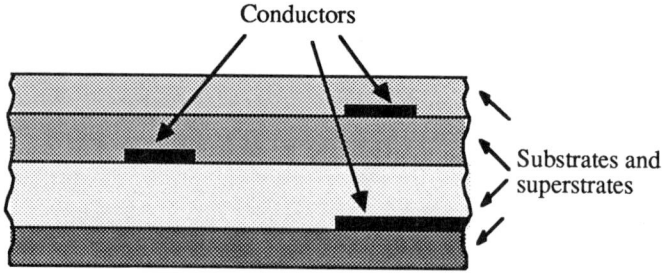

Fig. 1.6 Cross-section of a planar structure with thick conductors.

References

1. Pucel R.A., "Design consideration for monolithic microwave circuits", *IEEE Trans. Microwave Theory Tech.*, **MTT-29**, pp.513-532, 1981.

2. Alexopoulos N.G., "Integrated-circuit structures on anisotropic substrates", *ibid*, **MTT-33**, pp.847-881, 1985.

3. Baht B. and Koul S.K., *Analysis, Design and Applications of Finlines*, Artech House, Norwood, MA02062, 1987.

4. Chan C.H. and Mittra R., "Analysis of a class of cylindrical multiconductor transmission lines using an iterative approach", *IEEE Trans. Microwave Theory Tech.*, **MTT-35**, pp.415-423, 1987.

5. Alexopoulos N.G. and Nakatani A., "Cylindrical substrate microstrip line characterisation", *ibid*, pp.843-849, 1987.

6. Nakatani A. and Alexopoulos N.G., "Coupled microstrip lines on a cylindrical substrate", *ibid*, pp.1393-1398, 1987.

7. Itoh T. and Mittra R., "Spectral domain approach for calculating the dispersion characteristics of microstrip lines", *ibid*, **MTT-21**, pp.496-499, 1973.

8. Jansen R.H., Arnold R.G. and Eddison I.G., "A comprehensive CAD approach to the design of MMICs up to mm-wave frequencies", *ibid*, **MTT-36**, pp.208-218, 1988.

9. Jansen R.H., "A novel CAD tool and concept compatible with the requirements of multilayer GaAs MMIC technology", *IEEE*, **MTT-S Digest**, pp.711-714, 1985.

10. Pramanick P. and Bhartia P., "Accurate analysis equations and synthesis technique for unilateral finlines", *IEEE Trans. Microwave Theory Tech.*, **MTT-33**, pp.24-30, 1985.

11. Yamashita E., Atsuki K. and Ueda T., "An approximate dispersion formula of microstrip lines for computer-aided design of microwave integrated circuits", *ibid*, **MTT-27**, pp.1036-1038, 1979.

12. Jansen R.H., "Shielded rectangular microstrip disc resonators", *Electron. Lett.*, **Vol. 10**, pp.299-300, 1974.

13. Jansen R.H., "Computer analysis of edge-coupled planar structures", *ibid*, pp.520-522, 1974.

14. Rahmat-Samii Y., Itoh T. and Mittra R., "A spectral domain analysis for solving microstrip discontinuity problems", *IEEE Trans. Microwave Theory Tech.*, **MTT-22**, pp.372-378, 1974.

15. Itoh T. and Mittra R., "A technique for computing dispersion characteristics of shielded microstrip lines", *ibid*, pp.896-898, 1974.

16. Davies J.B. and Mirshekar-Syahkal D., "Spectral domain solution of arbitrary coplanar transmission line with multilayer substrate", *ibid*, **MTT-25**, pp.143-146, 1977.

17. Itoh T., "Spectral domain immittance approach for dispersion characteristics of generalized printed transmission lines", *ibid*, **MTT-28**, pp.733-736, 1980.

18. Itoh, T., ed., "*Planar Transmission Line Structures*, IEEE Press, New York, 1987.

Chapter 2

BASIC EQUATIONS AND CONCEPTS

This chapter briefly reviews some of the basic material in the electromagnetic theory which will be frequently referred to in the following chapters. The topics included are Maxwell's equations, constitutive relations, boundary conditions, edge conditions and Hertzian potential functions.

2.1 Maxwell's Equations

It is well-established that classical electromagnetic phenomena on a macroscopic scale can be explained or predicted by Maxwell's equations. In differential form, these equations are:

$$\nabla \times \mathbf{E}(t) = -\frac{\partial \mathbf{B}(t)}{\partial t} \tag{2.1.a}$$

$$\nabla \times \mathbf{H}(t) = \frac{\partial \mathbf{D}(t)}{\partial t} + \mathbf{J} \tag{2.1.b}$$

$$\nabla \cdot \mathbf{D}(t) = \rho(t) \tag{2.1.c}$$

$$\nabla \cdot \mathbf{B}(t) = 0 \tag{2.1.d}$$

where $\mathbf{E}(t)$ and $\mathbf{D}(t)$ represent electric field and flux density, $\mathbf{H}(t)$ and $\mathbf{B}(t)$ denote magnetic field and flux density, $\mathbf{J}(t)$ and $\rho(t)$ show electric current density and electric charge density.

For time harmonic electromagnetic fields with the angular frequency ω, Maxwell's equations assume a simple form given by:

CHAP. 2 BASIC EQUATIONS AND CONCEPTS

$$\nabla \times E = -j\omega B \quad (2.2.a)$$

$$\nabla \times H = j\omega D + J \quad (2.2.b)$$

$$\nabla \cdot D = \rho \quad (2.2.c)$$

$$\nabla \cdot B = 0 \quad (2.2.d)$$

where E, D, H, B, J and ρ are the time-independent forms of $E(t)$, $D(t)$, $H(t)$, $B(t)$, $J(t)$ and $\rho(t)$ respectively. Rigorously speaking, $E(t) = E \, exp \, (j\omega t)$, $H(t) = H \, exp \, (j\omega t)$, and so on. In this book we restrict ourselves to time-harmonic fields.

2.2 Constitutive Relations

The role of medium enters Maxwell's equations through a set of equations called constitutive relations:

$$D = \varepsilon \cdot E \quad (2.3.a)$$

$$B = \mu \cdot H \quad (2.3.b)$$

$$J = \sigma \cdot E \quad (2.3.c)$$

where ε (permittivity), μ (permeability) and σ (conductivity) are the electrical properties of the medium. Although generally these parameters are described by second rank tensors (dyadics), one, two or all of these parameters are scalar quantities for the commonest media.

Unless otherwise specified, it is assumed in this book that μ, ε and σ are scalar and are simply shown by μ, ε and σ.

2.3 Boundary Conditions

At the boundary of two different media, the electromagnetic field satisfies a set of equations called the boundary conditions. These equations, in effect, provide a facility for matching the field at a boundary.

The general forms of boundary conditions are as follows [1]:

CHAP. 2 BASIC EQUATIONS AND CONCEPTS

$$n \times (E_2 - E_1) = 0 \tag{2.4.a}$$

$$n \times (H_2 - H_1) = J_s \tag{2.4.b}$$

$$n \cdot (D_2 - D_1) = \rho_s \tag{2.4.c}$$

$$n \cdot (B_2 - B_1) = 0 \tag{2.4.d}$$

where subscripts *1* and *2* specify the media and n indicates the normal unit vector at the boundary, Fig. 2.1. In the above equation J_s and ρ_s are the surface current density and the surface charge density respectively, which may exist at the interface of the two media.

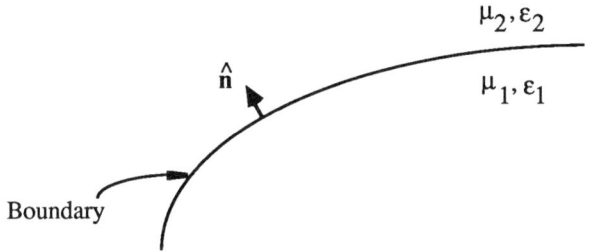

Fig. 2.1

In an electromagnetic problem involving two or several media, Maxwell's equations together with the boundary conditions are required to be simultaneously satisfied. This ensures that the field behaviour is uniquely determined. From the boundary conditions, Eqs. (2.4), it is not difficult to deduce that in order to match the field at the interface of two media, six equations should be satisfied. However, given that the fields are solutions of Maxwell's equations in each medium, two of these six equations are redundant. For instance, if the tangential components of the fields are matched, Eqs. (2.4.a) and (2.4.b), the normal components of D and B will automatically satisfy Eqs. (2.4.c) and (2.4.d). The proof of this can be found in [1].

As an example, let us write down the boundary conditions at the air-dielectric interface of the planar structure shown in Fig. 2.2. This figure illustrates the cross-section of an idealized microstrip line where the strip is assumed to be infinitely conductive ($\sigma = \infty$), and to have negligible thickness. For the given interface, the explicit forms of the boundary conditions for the electric field components are:

$$E_{x,2} - E_{x,1} = 0 \qquad 0 \le |x| < \infty \qquad (2.5.a)$$

$$\varepsilon_2 E_{y,2} - \varepsilon_1 E_{y,1} = \begin{cases} 0 & w \le |x| < \infty \\ \rho_s & 0 \le |x| < w \end{cases} \qquad (2.5.b)$$

$$E_{z,2} - E_{z,1} = 0 \qquad 0 \le |x| < \infty \qquad (2.5.c)$$

and for the magnetic field components are:

$$H_{x,2} - H_{x,1} = \begin{cases} 0 & w \le |x| < \infty \\ -J_z & 0 \le |x| < w \end{cases} \qquad (2.5.d)$$

$$H_{y,2} - H_{y,1} = 0 \qquad 0 \le |x| < \infty \qquad (2.5.e)$$

$$H_{z,2} - H_{z,1} = \begin{cases} 0 & w \le |x| < \infty \\ J_x & 0 \le |x| < w \end{cases} \qquad (2.5.f)$$

As can be seen from the above equations, the presence of strip between two dielectric media, Fig. 2.2, leads to complicated boundary conditions with the consequence of making the field solution difficult to achieve.

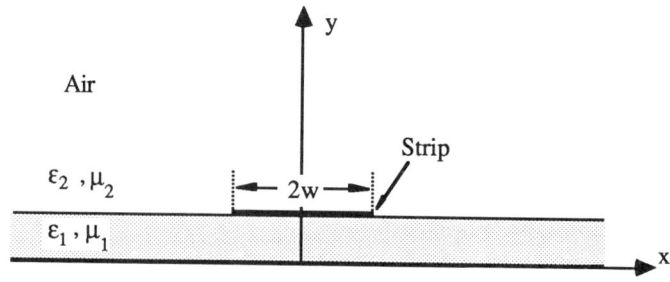

Fig. 2.2

2.4 Edge Condition

In an electromagnetic problem involving sharp edges, some of the field components can have unbounded behaviour in the vicinity of these edges. The field singularities are, however, such that the physical field solution (the unique solution) of the problem

CHAP. 2 BASIC EQUATIONS AND CONCEPTS

satisfies the edge condition. This condition calls for the energy function

$$W = \frac{1}{4}\int_V (\varepsilon|E|^2 + \mu|H|^2)dv \qquad (2.6)$$

to remain bounded within an infinitesimal volume V surrounding the edge.

There are many instances in the solutions of electromagnetic problems where a prior knowledge of the field singularities near sharp edges helps to reduce the mathematical labour. Particularly in modern electromagnetic problems where the majority of solutions are sought through numerical techniques, implementation of these singularities in the solution leads to some favourable features. These include enhancement of the accuracy of the numerical solution [2,3], feasibility of avoiding an undesirable effect called the relative convergence, (ie: converging to nonphysical solutions) [4,5] and achievement of high computing efficiency.

A valuable study by Meixner [6] has been the basis for many investigations carried out to assess the field distribution at the edge of a wedge surrounded by several media, Fig. 2.3 [7,8,9]. These studies are not to be discussed here. Instead, in the context of this book, it is sufficient to acknowledge the field singularities at the edges of 0° and 90° conducting wedges enclosed in two homogeneous media, Fig. 2.4. These singularities can be obtained by applying the technique presented in [6,10] to an arbitrary wedge surrounded by two dielectrics as shown in Fig. 2.5. Results of the

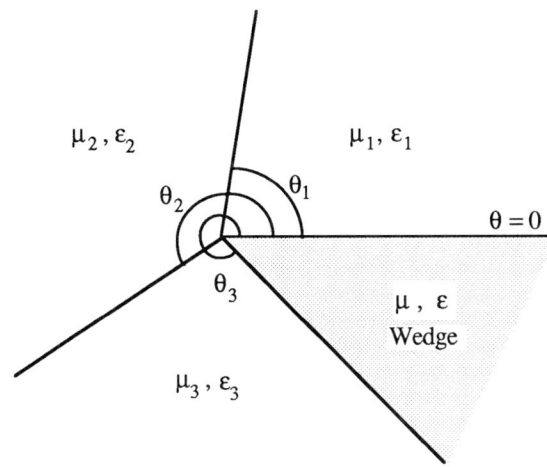

Fig. 2.3 A dielectric wedge surrounded by three other media.

CHAP. 2 BASIC EQUATIONS AND CONCEPTS

application of this technique are as follows. At the edge of the wedge $(r \to 0)$, the electromagnetic field can be described by a combination of two sets of asymptotic solutions

$$H_z, E \to r^p \qquad (2.7.a)$$

$$H_t \to r^{-1+p} \qquad (2.7.b)$$

and

$$E_t \to r^{-1+p} \qquad (2.8.a)$$

$$E_z, H \to r^p \qquad (2.8.b)$$

where p for the first set is found from the equation

$$(1 - \frac{\mu_2}{\mu_1})\cos(p\theta_2) + [\tan(p\theta_1) + \frac{\mu_2}{\mu_1}\cot(p\theta_1)]\sin(p\theta_2) = 0 \qquad (2.9)$$

and for the second set from the equation

$$(1 - \frac{\varepsilon_2}{\varepsilon_1})\cos(p\theta_2) - [\cot(p\theta_1) + \frac{\varepsilon_2}{\varepsilon_1}\tan(p\theta_1)]\sin(p\theta_2) = 0 \qquad (2.10)$$

From the above equations, it is clear that for each of the above sets, p can have multiple solution. The admissible field components are, however, those with dominant singularities. Note that using Eq. (2.6), it is easy to check that $p > 0$.

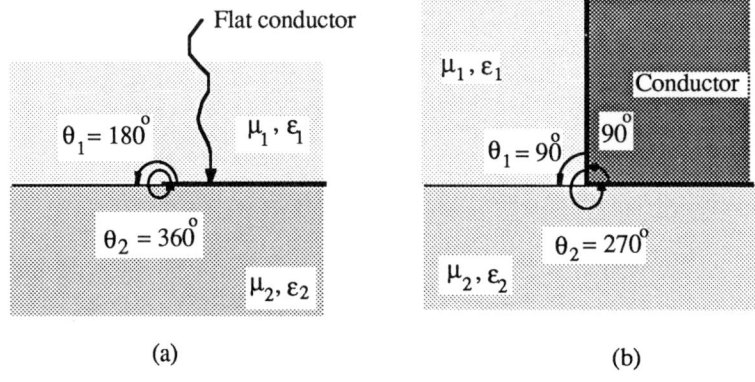

Fig. 2.4 Two types of edges encountered in planar structures, (a) 0° edge, (b) 90° edge.

CHAP. 2 BASIC EQUATIONS AND CONCEPTS 17

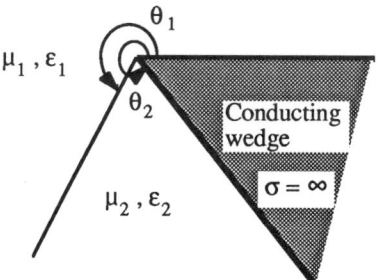

Fig. 2.5 A conducting wedge embeded in two media.

As an example, let us find the behaviour of the field near the edge of a flat conducting plate shown in Fig. 2.6. We start off with Eqs. (2.9) and (2.10); in order to find $p's$ we first substitute for θ_1 and θ_2 in these equations to obtain:

$$\sin 2\pi p = 0 \tag{2.11}$$

$$(1 - \frac{1}{\varepsilon_r})\cos 2\pi p - (\cot \frac{\pi}{2} p + \frac{1}{\varepsilon_r}\tan \frac{\pi}{2} p)\sin 2\pi p = 0 \tag{2.12}$$

Solutions to Eq. (2.11) are

$$p = 0, \tfrac{1}{2}, 1, \ldots \tag{2.13}$$

of which $p = \tfrac{1}{2}$ is the only acceptable solution. Thus the first set of solutions, Eq.

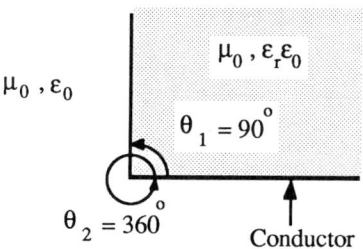

Fig. 2.6 A plane sheet of perfect conductor surrounded by two dielectrics.

CHAP. 2 BASIC EQUATIONS AND CONCEPTS

(2.7), is

$$H_z, E \to r^{\frac{1}{2}} \quad (2.14.a)$$

$$H_t \to r^{-\frac{1}{2}} \quad (2.14.b)$$

Solutions to Eq. (2.12) are

$$p = k + \frac{1}{2} + \frac{1}{\pi}\sin^{-1}\frac{\varepsilon_r - 1}{2(\varepsilon_r + 1)} \quad (2.15)$$

where

$$k = 0, 1, 2, \ldots \quad (2.16)$$

of which $k = 0$ produces the acceptable solution:

$$p = \frac{1}{2} + \frac{1}{\pi}\sin^{-1}\frac{\varepsilon_r - 1}{2(\varepsilon_r + 1)} \quad (2.17)$$

Hence, from the second set, Eq. (2.8), we obtain

$$E_t \to r^{-\frac{1}{2} + q} \quad (2.18.a)$$

$$E_z, H_t \to r^{\frac{1}{2} + q} \quad (2.18.b)$$

where

$$q = \frac{1}{\pi}\sin^{-1}\frac{\varepsilon_r - 1}{2(\varepsilon_r + 1)} \quad (2.19)$$

From (2.14) and (2.18), the dominant components are

$$E_t \to r^{-\frac{1}{2} + q} \quad (2.20.a)$$

$$H_t \to r^{-\frac{1}{2}} \quad (2.20.b)$$

$$E_z, H_z \to r^{\frac{1}{2}} \quad (2.20.c)$$

which represent the field behaviour near the edge.

In modelling planar structures, it is very common to assume that strips have no thickness. Using the technique presented in the above example, it is easy to show that the field near the edges of such strips sandwiched between two dielectrics ($\mu_1 = \mu_2$), Fig. 2.4.a, behaves as

CHAP. 2 BASIC EQUATIONS AND CONCEPTS

$$E_t, H_t \to r^{-\frac{1}{2}} \qquad (2.21.a)$$

$$E_z, H_z \to r^{\frac{1}{2}} \qquad (2.21.b)$$

2.5 Potential Functions

Using Maxwell's equations, it is possible to show that in a source-free medium, the electromagnetic field can be derived from either an electric vector potential function Φ,

$$\mathbf{E} = \nabla \times \nabla \times \Phi \qquad (2.22.a)$$

$$\mathbf{H} = j\omega\varepsilon \nabla \times \Phi \qquad (2.22.b)$$

or from a magnetic vector potential function Ψ

$$\mathbf{E} = -j\omega\mu \nabla \times \Psi \qquad (2.23.a)$$

$$\mathbf{H} = \nabla \times \nabla \times \Psi \qquad (2.23.b)$$

The two functions Φ and Ψ which are known as Hertzian vector potential functions are solutions of the homogeneous vector Helmholtz equation (wave equation) [1].

$$\nabla^2 U + \omega^2 \mu\varepsilon U = 0 \qquad (2.24)$$

An application of the above expressions can be found [1] in the theory of cylindrical waveguides, Fig. 2.7. Fields associated with TM and TE modes may be derived from

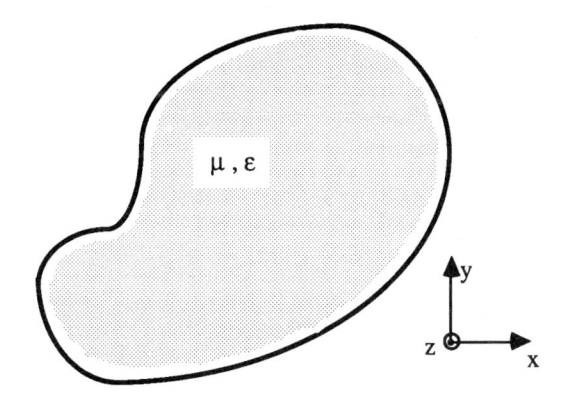

Fig. 2.7 Cross-section of a cylindrical waveguide.

the scalar electric potential function $\phi(x,y)$ and from the scalar magnetic potential function $\psi(x,y)$ respectively. These functions are related to Φ and Ψ as follows:

$$\Phi = \phi(x,y) e^{-j\beta z} k \qquad (2.25)$$

$$\Psi = \psi(x,y) e^{-j\beta z} k \qquad (2.26)$$

and are solutions of the scalar Helmholtz equation:

$$\nabla^2 U(x,y) + (\omega^2 \mu \varepsilon - \beta^2) U(x,y) = 0 \qquad (2.27)$$

In the above equations β denotes the propagation constant and k denotes the unit vector in the z direction. From Eqs. (2.22) and (2.23) the field component for TM modes are:

$$E_t = -j\beta \nabla_t \phi(x,y) e^{-j\beta z} \qquad (2.28.a)$$

$$H_t = -j\omega\varepsilon k \times \nabla_t \phi(x,y) e^{-j\beta z} \qquad (2.28.b)$$

$$E_z = (\omega^2 \mu \varepsilon - \beta^2) \phi(x,y) e^{-j\beta z} \qquad (2.28.c)$$

and for TE modes are

$$E_t = j\omega\mu k \times \nabla_t \psi(x,y) e^{-j\beta z} \qquad (2.29.a)$$

$$H_t = -j\beta \nabla_t \psi(x,y) e^{-j\beta z} \qquad (2.29.b)$$

$$H_z = (\omega^2 \mu \varepsilon - \beta^2) \psi(x,y) e^{-j\beta z} \qquad (2.29.c)$$

where

$$\nabla_t = \frac{\partial}{\partial x} i + \frac{\partial}{\partial y} j \qquad (2.30)$$

and i and j are the x and y directed unit vectors.

Now consider the more general case of cylindrical waveguides loaded with several dielectrics, Fig. 2.8. In these waveguides no TE or TM modes can essentially exist individually [1]. The associated modes are hybrid which can be assumed as a combination of TE and TM modes. Therefore, in view of Eqs. (2.28) and (2.29), the field components in the i th dielectric can be expressed as:

CHAP. 2 BASIC EQUATIONS AND CONCEPTS   21

$$E_{z,i} = (\omega^2 \mu_i \varepsilon_i - \beta^2) \phi_i e^{-j\beta z} \qquad (2.31.a)$$

$$H_{z,i} = (\omega^2 \mu_i \varepsilon_i - \beta^2) \psi_i e^{-j\beta z} \qquad (2.31.b)$$

$$E_{t,i} = -j\beta (\nabla_t \phi_i - \frac{\omega \mu_i}{\beta} k \times \nabla_t \psi_i) e^{-j\beta z} \qquad (2.31.c)$$

$$H_{t,i} = -j\beta (\nabla_t \psi_i + \frac{\omega \varepsilon_i}{\beta} k \times \nabla_t \phi_i) e^{-j\beta z} \qquad (2.31.d)$$

In the above equations, $\phi_i = \phi_i(x,y)$ and $\psi_i = \psi_i(x,y)$ are the scalar electric and magnetic potential functions associated with the *i th* dielectric region and, like $\phi(x,y)$ and $\psi(x,y)$, they are solutions of the wave equation:

$$\nabla_t^2 u_i + (\omega^2 \mu_i \varepsilon_i - \beta^2) u_i = 0 \qquad (2.32)$$

Although the potential functions referred to above are the most common types of potential functions in the waveguide theory, they are by no means the only possibilities. Depending on the waveguide structure, a better choice of potential functions may exist, leading to a simpler description of the waveguide field. For instance, in some dielectric loaded rectangular waveguides, although both E_z and H_z are present, one of the transverse components of the field normal to the dielectric interface is zero [1]. In these cases, it is more convenient to write all the field components in terms of the nonzero component normal to the interface. This

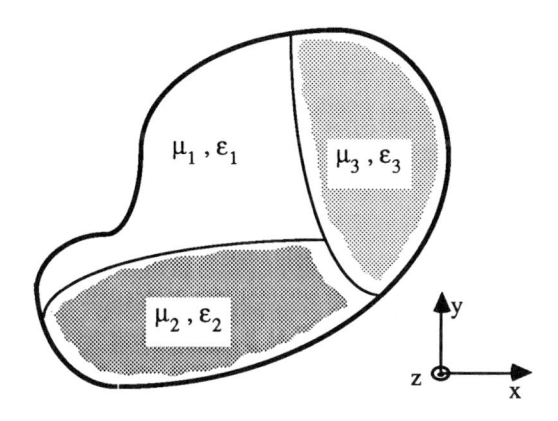

Fig. 2.8 Cross-section of a cylindrical waveguide loaded with several dielectrics.

component, in fact, plays the role of a potential function. Such field representation has given rise to the introduction of a set of modes known as the LSE (longitudinal section electric) mode and the LSM (longitudinal section magnetic) mode [1]. As pointed out earlier, a hybrid mode can be expressed as a combination of TE and TM modes. It is obvious that it can also be described in terms of LSE and LSM modes. Of course, depending on the strategy taken for the solution of an inhomogeneously filled waveguide, one of the two representations of the hybrid mode is preferable. This matter will be clarified further, later in the book.

References

1. Collin R., *Field Theory of Guided Waves*, McGraw-Hill, New York, 1960.

2. Pantic-Tanner Z., Chan C.H. and Mittra R., "The treatment of edge singularities in the full-wave finite element solution of waveguiding problems", North American Radio Science Meeting, Syracuse, N.Y., 1988.

3. Pantic Z. and Mittra R., "Quasi-TEM analysis of microwave transmission lines by the finite element method", *IEEE Trans. Microwave Theory Tech.*, **MTT-34**, pp.1096-1103, 1986.

4. Mittra R. "Relative convergence of the solution of a doubly infinite set of equations"; *J. Res. NBS*, **Vol. 670**, pp.245-254, 1963.

5. Lee S.W., Jones W.R. and Campbell J.J., "Convergence of numerical solutions of iris-type discontinuity problems", *IEEE Trans. Microwave Theory Tech.*, **MTT-19**, pp.528-536, 1971.

6. Meixner J., "The behaviour of electromagnetic fields at edges", *IEEE Trans. Antennas and Propagation*, **AP-20**, pp.442-446, 1972.

7. Hurd R.A., "The edge condition of electromagnetics", *ibid*, **AP-24**, pp.70-73, 1976.

8. Brooke G.H. and Kharadly M.M., "Field behaviour near anisotropic and multi-dielectric edges", *ibid*, **AP-25**, pp.571-575, 1977.

9. Chan C.H., Pantic-Tanner Z. and Mittra R. "Field behaviour near a conducting edge embedded in an inhomogeneous anisotropic medium", *Electron. Lett.*, **Vol.24**, No.6, pp.355-356, 1988.

10. Mittra R. and Lee S.W., *Analytical Techniques in the Theory of Guided Waves*, Macmillan, New York, 1971.

Chapter 3

INTRODUCTION TO THE SPECTRAL DOMAIN TECHNIQUE

In this chapter, the spectral domain technique is introduced. This is accomplished through the spectral domain formulation of the most basic planar structure, the microstrip line. The solution of this example allows the basic concepts behind the spectral domain technique to be easily exposed. Also mathematical principles used in the development of the spectral domain technique will be briefly discussed in this chapter.

3.1 Microstrip Line and the Spectral Domain Technique

Before proceeding with the formal treatment of the microstrip line by the spectral domain technique, it seems necessary to devote a few lines to the physical structure of this transmission line.

It is rather common knowledge now that a microstrip line consists of a substrate, a ground plane and a metal strip etched on the top of the substrate, Fig. 3.1. In the microstrip line, conductors are usually gold plated and the substrate material, depending on the application requirements, can be a dielectric, a ferrite, a semiconductor or something else. As part of a circuit, the microstrip line together with other circuit components are enclosed within a metal box. The function of the box is to provide the electromagnetic shielding as well as to protect the circuit against shocks and vibrations. In studying microstrip lines, the size and the shape of the box are important factors. However, it is usually possible with some approximation to use one of the models in Fig. 3.2 for the analysis of a microstrip line.

Each of the microstrip models presented in Fig. 3.2 can accept the spectral domain solution [1,2] provided that the strip is assumed to be negligibly thin ($t \approx 0$) and

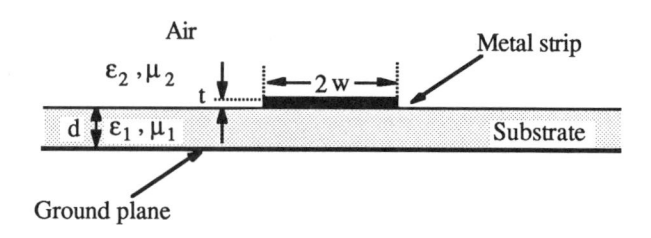

Fig. 3.1 Cross-section of the microstrip line.

the metals involved are assumed to be perfect conductors. For the majority of the microstrip lines, these are legitimate assumptions, particularly the first one when we consider that in practice t is usually much less than the substrate thickness d. The substrates in the given models can be anisotropic and/or lossy. There are ample works reported in the literature describing the use of the spectral domain technique in analysing microstrip lines with such substrates; for instance, see references [3-11]. Some of these works will be explained later in the book.

In the following sections, we make use of the model shown in Fig. 3.2.c to introduce the spectral domain technique. This model is more general because by letting $a \to \infty$ and/or $h \to \infty$, the two other microstrip models can be simulated. However, since modes associated with closed guides are discrete [12], the spectral domain solution of the shielded microstrip line when a and/or h tends to infinity will only render the discrete modes of the open microstrip line [1].

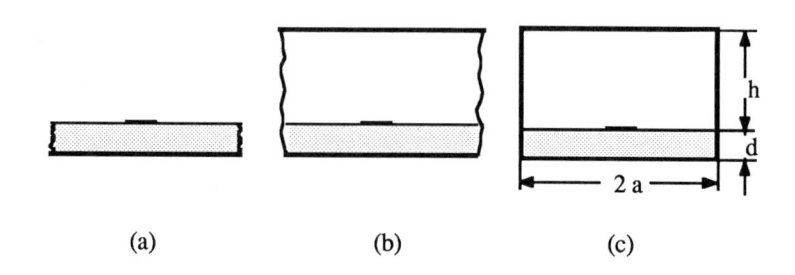

Fig. 3.2 Cross-section of (a) an open microstrip line, (b) a laterally open microstrip line and (c) a shielded microstrip line.

3.2 Spectral Domain Formulation of Microstrip Lines

As pointed out in the above section, in the following study we consider a microstrip line with the cross-section shown in Fig. 3.2.c which is fully specified in Fig. 3.3. Suppose that this line is driven by a sinusoidal source of angular frequency ω and assume that the objective is to find an expression which gives the dispersion characteristics of this line.

It is clear that the dispersion expression can be found upon the successful solution of Maxwell's equations, Eqs. (2.2), subject to boundary conditions for the structure of concern. In the present problem, we are dealing with a guide which is uniform along the direction of propagation. Therefore the notion of scalar potential functions presented in section (2.5) can be borrowed to formulate the problem. Since the microstrip line can only support hybrid modes (TE and TM modes simultaneously) [13], two scalar potential functions, ϕ_i and ψ_i, are required in order to obtain the full-wave solution. Subscript $i = 1, 2$ in these functions indicate the two dielectric media in Fig. 3.3. We remember that ϕ_i and ψ_i are functions of x and y and are solutions of the wave equation:

$$\nabla_t^2 u_i + (k_i^2 - \beta^2) u_i = 0 \tag{3.1}$$

where

$$k_i^2 = \omega^2 \mu_i \varepsilon_i \tag{3.2}$$

and β denotes the phase constant of the wave in the z-direction.

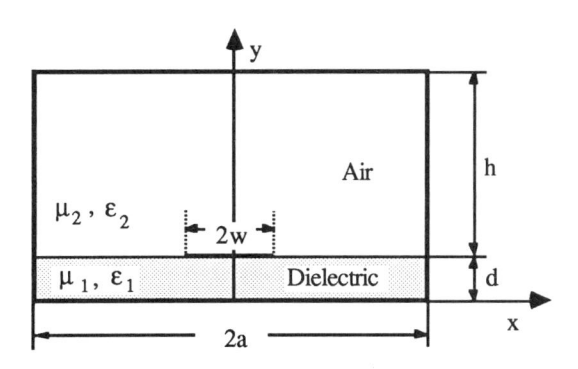

Fig. 3.3 Cross-section of the microstrip line model considered in the analysis.

3.2.1 Potential Functions in the Fourier Domain

In the spectral domain technique, we often encounter terms such as potential functions or field components in the Fourier domain. These concepts can be developed as follows.

Consider the two scalar potential functions ϕ_i and ψ_i introduced in the previous section. The Fourier series theory [14-16, Appendix I] allows these functions to be expanded in terms of trigonometric functions of x as follows:

$$\phi_i = \sum_{n=-\infty}^{+\infty} \tilde{\phi}_i(\alpha_n, y) e^{-j\alpha_n x} \qquad (3.3.a)$$

$$\psi_i = \sum_{n=-\infty}^{+\infty} \tilde{\psi}_i(\alpha_n, y) e^{-j\alpha_n x} \qquad (3.3.b)$$

In the above expansions, α_n is the Fourier parameter and is found by examining the field behaviour along the x-direction. Since the field components $E_{z,i}$ and $H_{z,i}$ are proportional to ϕ_i and ψ_i respectively (see Eqs. (2.31)), examination of these two field components for determination of α_n is sufficient. For instance, in the present problem, if we concern ourselves with the even modes of the microstrip line only, then:

$$E_{z,i}(x,y,z) = E_{z,i}(-x,y,z) \qquad (3.4.a)$$

$$H_{z,i}(x,y,z) = -H_{z,i}(-x,y,z) \qquad (3.4.b)$$

The above equations and the boundary conditions at the two side walls are satisfied when

$$\alpha_n = (n+\tfrac{1}{2})\frac{\pi}{a} \qquad (3.5)$$

Let us check this, for instance, for $E_{z,i}$. From Eq. (2.31.a) and Eq. (3.3.a), the following expression can be obtained,

$$E_{z,i} = (k_i^2 - \beta^2)\left[\sum_{n=0}^{\infty} \tilde{\phi}_i(\alpha_n, y) e^{-j\alpha_n x} + \sum_{n=-\infty}^{0} \tilde{\phi}_i(\alpha_n, y) e^{-j\alpha_n x}\right] e^{-j\beta z} \qquad (3.6)$$

Since $E_{z,i}$ is an even function of x, $\tilde{\phi}_i(\alpha_n, y)$ will be an even function of α_n. Thus, the second term on the right-hand side of Eq. (3.6) can be replaced by

CHAP. 3 INTRODUCTION TO THE SPECTRAL DOMAIN TECHNIQUE

$$\sum_{n=0}^{\infty} \tilde{\phi}_i(\alpha_n, y) e^{j\alpha_n x} \tag{3.7}$$

Hence, the alternative form of Eq. (3.6) is

$$E_{z,i} = 2(k^2 - \beta^2) \sum_{n=0}^{\infty} \tilde{\phi}_i(\alpha_n, y) \cos \alpha_n x \; e^{-j\beta z} \tag{3.8}$$

From the above equation, it is clear that E_z is an even function of x and for the given α_n, Eq. (3.5), it meets the boundary conditions at the side walls ($x = \pm a$). Note that in the case of odd modes, α_n is given by

$$\alpha_n = \frac{n\pi}{a} \tag{3.9}$$

Let us turn to Eqs. (3.3.a) and (3.3.b) now and define an integral transformation between ϕ_i, ψ_i and their Fourier transform coefficients. Since these equations involve Fourier series, the integral transformation relating the potential functions to their respective spectra is naturally the Fourier integral [14-16]. Thus

$$\tilde{\phi}_i(\alpha_n, y) = \frac{1}{2a} \int_{-a}^{+a} \phi_i e^{j\alpha_n x} dx \tag{3.10.a}$$

$$\tilde{\psi}_i(\alpha_n, y) = \frac{1}{2a} \int_{-a}^{+a} \psi_i e^{j\alpha_n x} dx \tag{3.10.b}$$

where the left-hand sides of (3.10.a) and (3.10.b) are the finite Fourier transforms of ϕ_i and ψ_i respectively.

From now on, we show the finite Fourier transform of a function with period $2a$ as (see Appendix I)

$$\tilde{\xi}(\alpha_n) = \frac{1}{2a} \int_{-a}^{+a} \xi e^{j\alpha_n x} dx \tag{3.11.a}$$

or in short notation as

$$\tilde{\xi} = F(\xi) \tag{3.11.b}$$

In this book, the above notation is also used to represent the conventional Fourier transform [14-16] where $a \to \infty$, ie:

$$\tilde{\xi}(\alpha) = \frac{1}{2\pi} \int_{-\infty}^{+\infty} \xi e^{j\alpha x} dx \qquad (3.11.c)$$

The Fourier transform defined by Eqs. (3.10) or Eqs. (3.11) can be applied to the wave equation, Eq. (3.1), as well. The result is as follows:

$$\frac{d^2}{dy^2} \tilde{u}_i - \gamma_{i,n}^2 \tilde{u}_i = 0 \qquad (3.12)$$

where

$$\gamma_{i,n}^2 = \alpha_n^2 + \beta^2 - k_i^2 \qquad (3.13)$$

In the derivation of the above equation, you may require to know that

$$F(\frac{\partial}{\partial x} u_i) = -j\alpha_n \tilde{u}_i \qquad (3.14.a)$$

and

$$F(\frac{\partial^2}{\partial x^2} u_i) = -\alpha_n^2 \tilde{u}_i \qquad (3.14.b)$$

The general solution to Eq. (3.12) is

$$\tilde{u}_i = R_{i,n} \sinh(\gamma_{i,n} y) + Q_{i,n} \cosh(\gamma_{i,n} y) \qquad (3.15)$$

where $R_{i,n}$ and $Q_{i,n}$ are constants to be defined by the boundary conditions.

As u_i denotes ϕ_i and ψ_i, in view of Eq. (3.15), the solutions for these potential functions in the Fourier domain are as follows:

$$\tilde{\phi}_1(\alpha_n, y) = A_{1,n} \sinh(\gamma_{1,n} y) \qquad (3.16.a)$$
$$\tilde{\psi}_1(\alpha_n, y) = C_{1,n} \cosh(\gamma_{1,n} y) \qquad (3.16.b)$$

$0 < y < d$

$$\tilde{\phi}_2(\alpha_n, y) = B_{2,n} \sinh\left[\gamma_{2,n}(h+d-y)\right] \qquad (3.17.a)$$
$$\tilde{\psi}_2(\alpha_n, y) = D_{2,n} \cosh\left[\gamma_{2,n}(h+d-y)\right] \qquad (3.17.b)$$

$d < y < h+d$

Note that the above equations satisfy the appropriate boundary conditions at the bottom and top walls of the shielded microstrip line, Fig. 3.3, ie:

CHAP. 3 INTRODUCTION TO THE SPECTRAL DOMAIN TECHNIQUE

$$E_{z,i} = 0 , \frac{\partial}{\partial y} H_{z,i} = 0 \quad \begin{cases} i = 1, y = 0 \\ i = 2, y = h+d \end{cases} \tag{3.18}$$

In equations (3.16) and (3.17), parameters $A_{1,n}$, $C_{1,n}$, $B_{2,n}$, $D_{2,n}$ and β (which is implicit in $\gamma_{i,n}$) are still unknown. As will be shown later, in order to determine these parameters, boundary conditions at the air-dielectric interface are to be used.

As may be appreciated by now, all that we have done so far is to develop an integral transformation in order to reduce the wave equation into a manageable equation in the transform domain (Fourier domain). Expressions (3.16) and (3.17) are solutions of the wave equation for the electric and magnetic potential functions in the Fourier domain. In view of Eqs. (2.31) and the Fourier transform defined in this section, it is therefore possible to express all the field components in the Fourier domain as follows:

$$\tilde{E}_{z,i} = (k_i^2 - \beta^2) \, \tilde{\phi}_i(\alpha_n, y) e^{-j\beta z} \tag{3.19.a}$$

$$\tilde{H}_{z,i} = (k_i^2 - \beta^2) \, \tilde{\psi}_i(\alpha_n, y) e^{-j\beta z} \tag{3.19.b}$$

$$\tilde{E}_{t,i} = -j\beta \left[\tilde{\nabla}_t \tilde{\phi}_i(\alpha_n, y) - \frac{\omega\mu_i}{\beta} k \times \tilde{\nabla}_t \tilde{\psi}_i(\alpha_n, y) \right] e^{-j\beta z} \tag{3.19.c}$$

$$\tilde{H}_{t,i} = -j\beta \left[\tilde{\nabla}_t \tilde{\psi}_i(\alpha_n, y) + \frac{\omega\varepsilon_i}{\beta} k \times \tilde{\nabla}_t \tilde{\phi}_i(\alpha_n, y) \right] e^{-j\beta z} \tag{3.19.d}$$

where

$$\tilde{\nabla}_t = -j\alpha_n i + \frac{\partial}{\partial y} j \tag{3.20}$$

and i and j are the x and y directed unit vectors.

3.2.2 Boundary Conditions in the Space and in the Fourier Domain

In the second chapter, section 2.3, we set out the six boundary conditions associated with the air-dielectric interface in the microstrip line. However, as stated there, if the field in each dielectric region is the solution of Maxwell's equations, then it is sufficient to match the tangential components of the fields at the dielectric interface in order to obtain the coefficients adhered to these solutions and hence to determine the global field behaviour in the microstrip line. At the boundary of concern ($|x|<a$, $y = d$), the relations between these field components are as follows:

$$E_{z,2} - E_{z,1} = 0 \qquad (3.21.a)$$

$$E_{x,2} - E_{x,1} = 0 \qquad (3.21.b)$$

$$H_{z,2} - H_{z,1} = J_x(x)e^{-j\beta z} \qquad (3.21.c)$$

$$H_{x,2} - H_{x,1} = -J_z(x)e^{-j\beta z} \qquad (3.21.d)$$

where surface current density functions $J_x(x)$ and $J_z(z)$ are non-zero on the strip only, ie:

$$\begin{aligned} J_x(x), J_z(x) &\neq 0 & 0<|x|<w \\ &= 0 & \text{otherwise} \end{aligned} \qquad (3.22)$$

Applying the Fourier transform, Eq. (3.11.a), to the above boundary conditions, results in the following equations in the spectral domain:

$$\tilde{E}_{z,2} - \tilde{E}_{z,1} = 0 \qquad (3.23.a)$$

$$\tilde{E}_{x,2} - \tilde{E}_{x,1} = 0 \qquad (3.23.b)$$

$$\tilde{H}_{z,2} - \tilde{H}_{z,1} = \tilde{J}_x e^{-j\beta z} \qquad (3.23.c)$$

$$\tilde{H}_{x,2} - \tilde{H}_{x,1} = -\tilde{J}_z e^{-j\beta z} \qquad (3.23.d)$$

Let us compare the above two sets of equations, Eqs. (3.21) and (3.23), and note the difference. In the second set where the boundary conditions are expressed in the Fourier domain, the spectra of the current functions extend over the entire domain, (the spectra are continuous when $a \to \infty$, whereas in the space domain, these functions, which are not continuous functions of x, have values over the strip only, see Eq. (3.22). As will be shown in the following sections, this property of boundary conditions in the Fourier domain allows a convenient expression of the field in terms of the Fourier transforms of J_x and J_z.

3.2.3 Final Equations in the Fourier Domain

In the spectral domain technique, and indeed in several other techniques of solving planar structures, the objective is to reduce the solution of the problem to the solution of a set of coupled integral equations [17-26]. In the spectral domain technique, these equations relate the tangential components of electric field at the interfaces where strips of conductor are laid, to the current distributions of the strips. In this technique, the integral equations can be initially generated in the Fourier domain. To demonstrate this, we continue the solution of the microstrip line.

Substituting Eqs. (3.16) and (3.17) in Eqs. (3.19) leads to the expressions for the field components in the Fourier domain. These expressions for the dielectric medium are:

$$\tilde{E}_{x,1} = -(\alpha_n \beta A_{1,n} + j\omega\mu_1\gamma_{1,n}C_{1,n})\sinh(\gamma_{1,n}y)e^{-j\beta z} \quad (3.24.\text{a})$$

$$\tilde{E}_{y,1} = -j(\gamma_{1,n}\beta A_{1,n} + j\omega\mu_1\alpha_n C_{1,n})\cosh(\gamma_{1,n}y)e^{-j\beta z} \quad (3.24.\text{b})$$

$$\tilde{E}_{z,1} = (k_1^2 - \beta^2)A_{1,n}\sinh(\gamma_{1,n}y)e^{-j\beta z} \quad (3.24.\text{c})$$

$$\tilde{H}_{x,1} = -(\alpha_n \beta C_{1,n} - j\omega\varepsilon_1\gamma_{1,n}A_{1,n})\cosh(\gamma_{1,n}y)e^{-j\beta z} \quad (3.24.\text{d})$$

$$\tilde{H}_{y,1} = -j(\gamma_{1,n}\beta C_{1,n} - j\omega\varepsilon_1\alpha_n A_{1,n})\sinh(\gamma_{1,n}y)e^{-j\beta z} \quad (3.24.\text{e})$$

$$\tilde{H}_{z,1} = (k_1^2 - \beta^2)C_{1,n}\cosh(\gamma_{1,n}y)e^{-j\beta z} \quad (3.24.\text{f})$$

and for the air medium are

$$\tilde{E}_{x,2} = -(\alpha_n \beta B_{2,n} - j\omega\mu_2\gamma_{2,n}D_{2,n})\sinh[\gamma_{2,n}(d+h-y)]e^{-j\beta z} \quad (3.25.\text{a})$$

$$\tilde{E}_{y,2} = j(\gamma_{2,n}\beta B_{2,n} - j\omega\mu_2\alpha_n D_{2,n})\cosh[\gamma_{2,n}(d+h-y)]e^{-j\beta z} \quad (3.25.\text{b})$$

$$\tilde{E}_{z,2} = (k_2^2 - \beta^2)B_{2,n}\sinh[\gamma_{2,n}(d+h-y)]e^{-j\beta z} \quad (3.25.\text{c})$$

$$\tilde{H}_{x,2} = -(\alpha_n \beta D_{2,n} + j\omega\varepsilon_2\gamma_{2,n}B_{2,n})\cosh[\gamma_{2,n}(d+h-y)]e^{-j\beta z} \quad (3.25.\text{d})$$

$$\tilde{H}_{y,2} = j(\gamma_{2,n}\beta D_{2,n} + j\omega\varepsilon_2\alpha_n B_{2,n})\sinh[\gamma_{2,n}(d+h-y)]e^{-j\beta z} \quad (3.25.\text{e})$$

$$\tilde{H}_{z,2} = (k_2^2 - \beta^2)D_{2,n}\cosh[\gamma_{2,n}(d+h-y)]e^{-j\beta z} \quad (3.25.\text{f})$$

CHAP. 3 INTRODUCTION TO THE SPECTRAL DOMAIN TECHNIQUE 32

From the above equations, we now substitute the appropriate field components in the boundary conditions given in the Fourier domain, Eqs. (3.23). The result is a relation between parameters $A_{1,n}$, $B_{2,n}$, $C_{1,n}$, $D_{2,n}$ and the components of current distribution J_x and J_z as follows:

$$[a] \begin{bmatrix} A_{1,n} \\ B_{2,n} \\ C_{1,n} \\ D_{2,n} \end{bmatrix} = \begin{bmatrix} 0 \\ 0 \\ -\tilde{J}_z \\ \tilde{J}_x \end{bmatrix} \quad (3.26)$$

where the elements of the 4 × 4 matrix $[a]$ are given by the following expressions:

$$a_{13} = a_{14} = a_{41} = a_{42} = 0$$
$$a_{11} = -(k_1^2 - \beta^2)\sinh(\gamma_{1,n} d) \qquad a_{43} = -(k_1^2 - \beta^2)\cosh(\gamma_{1,n} d)$$
$$a_{12} = (k_2^2 - \beta^2)\sinh(\gamma_{2,n} h) \qquad a_{44} = (k_2^2 - \beta^2)\cosh(\gamma_{2,n} h)$$
$$a_{31} = -j\omega\varepsilon_1 \gamma_{1,n}\cosh(\gamma_{1,n} d) \qquad a_{21} = \alpha_n \beta \sinh(\gamma_{1,n} d) \quad (3.27)$$
$$a_{32} = -j\omega\varepsilon_2 \gamma_{2,n}\cosh(\gamma_{2,n} h) \qquad a_{22} = -\alpha_n \beta \sinh(\gamma_{2,n} h)$$
$$a_{33} = \alpha_n \beta \cosh(\gamma_{1,n} d) \qquad a_{23} = j\omega\mu_1 \gamma_{1,n}\sinh(\gamma_{1,n} d)$$
$$a_{34} = -\alpha_n \beta \cosh(\gamma_{2,n} h) \qquad a_{24} = j\omega\mu_2 \gamma_{2,n}\sinh(\gamma_{2,n} h)$$

It is clear that Eq. (3.26) can be combined with Eqs. (3.24) and (3.25) to give all the field components in terms of the current distribution on the strip. Remember that the components of this current distribution $J_x(x)$ and $J_z(x)$ together with β are still unknown. These parameters can be determined using the final condition existing at the air-dielectric boundary, namely

$$\begin{cases} E_{x,1} = E_{x,2} = 0 \\ E_{z,1} = E_{z,2} = 0 \end{cases} \qquad \begin{array}{l} 0 < |x| < w \\ y = d \end{array} \qquad \begin{array}{l} (3.28.\text{a}) \\ (3.28.\text{b}) \end{array}$$

$$J_x(x) = J_z(x) = 0 \qquad a > |x| > w \qquad (3.28.\text{c})$$

CHAP. 3 INTRODUCTION TO THE SPECTRAL DOMAIN TECHNIQUE

In order to exploit the above conditions, we write down $E_{x,1}$ and $E_{z,1}$ in terms of J_x and J_z in the Fourier domain initially. From Eqs. (3.24.a) and (3.24.c), it is clear that for the above purpose, we need to find $A_{1,n}$ and $C_{1,n}$. From Eq. (3.26) the explicit expressions for $A_{1,n}$ and $C_{1,n}$ in terms of the Fourier transforms of J_x and J_z are as follows:

$$A_{1,n} = (a_x \tilde{J}_x + a_z \tilde{J}_z) \sinh(\gamma_{2,n} h) \cosh(\gamma_{2,n} h) \cosh(\gamma_{1,n} d) \qquad (3.29)$$

$$C_{1,n} = (c_x \tilde{J}_x + c_z \tilde{J}_z) \sinh(\gamma_{2,n} h) \cosh(\gamma_{2,n} h) \cosh(\gamma_{1,n} d) \qquad (3.30)$$

where

$$a_x = \frac{j\omega(k_2^2 - \beta^2)\alpha_n \beta}{\delta} [\mu_1 \gamma_{1,n} \tanh(\gamma_{1,n} d) + \mu_2 \gamma_{2,n} \tanh(\gamma_{2,n} h)] \qquad (3.31)$$

$$a_z = \frac{-j\omega(k_2^2 - \beta^2)}{\delta} [\mu_1 \gamma_{1,n} (k_2^2 - \beta^2) \tanh(\gamma_{1,n} d) + \mu_2 \gamma_{2,n} (k_1^2 - \beta^2) \tanh(\gamma_{2,n} h)] \qquad (3.32)$$

$$c_x = \frac{(k_2^2 - \beta^2)}{\delta} \alpha_n \beta (k_1^2 - k_2^2) \tanh(\gamma_{1,n} d) \qquad (3.33)$$

$$c_z = \frac{1}{\delta} [\alpha_n^2 \beta^2 (k_2^2 - k_1^2) \tanh(\gamma_{1,n} d) + \omega^2 \mu_2 \varepsilon_1 \gamma_{2,n} \gamma_{1,n} (k_2^2 - \beta^2) \tanh(\gamma_{2,n} h) + k_2^2 \gamma_{2,n}^2 (k_1^2 - \beta^2) \tanh(\gamma_{1,n} d)] \qquad (3.34)$$

In the above equation δ is the determinant of $[a]$ given by the expression:

$$\delta = k_2^2 (k_1^2 - \beta^2)(k_2^2 - \beta^2) [\gamma_{1,n} \tanh(\gamma_{1,n} d) + \frac{\varepsilon_1}{\varepsilon_2} \gamma_{2,n} \tanh(\gamma_{2,n} h)][\gamma_{1,n} \coth(\gamma_{1,n} d) + \frac{\mu_1}{\mu_2} \gamma_{2,n} \coth(\gamma_{2,n} h)] \sinh(\gamma_{2,n} h) \cosh(\gamma_{2,n} h) \sinh(\gamma_{1,n} d) \cosh(\gamma_{1,n} d) \qquad (3.35)$$

Thus, at the air-dielectric interface $(y=d)$, the tangential electric field components in the Fourier domain are:

$$\tilde{E}_z = G_{11}(\alpha_n, \beta) \tilde{J}_x + G_{12}(\alpha_n, \beta) \tilde{J}_z \qquad (3.36.a)$$

$$\tilde{E}_x = G_{21}(\alpha_n, \beta) \tilde{J}_x + G_{22}(\alpha_n, \beta) \tilde{J}_z \qquad (3.36.b)$$

where:

CHAP. 3 INTRODUCTION TO THE SPECTRAL DOMAIN TECHNIQUE

$$G_{11}(\alpha_n,\beta) = G_{22}(\alpha_n,\beta) = \frac{j\omega}{k_2^2} \{\alpha_n\beta [\mu_1\gamma_{1,n}\tanh(\gamma_{1,n}d) \quad (3.37.a)$$
$$+ \mu_2\gamma_{2,n}\tanh(\gamma_{2,n}h)]\}/\Delta$$

$$G_{12}(\alpha_n,\beta) = -\frac{j\omega}{k_2^2} [\mu_1\gamma_{1,n}(k_2^2-\beta^2)\tanh(\gamma_{1,n}d) \quad (3.37.b)$$
$$+ \mu_2\gamma_{2,n}(k_1^2-\beta^2)\tanh(\gamma_{2,n}h)]/\Delta$$

$$G_{21}(\alpha_n,\beta) = -\frac{j\omega}{k_2^2} [\mu_1\gamma_{1,n}(k_1^2-\alpha_n^2)\tanh(\gamma_{1,n}d) \quad (3.37.c)$$
$$+ \mu_2\gamma_{2,n}(k_2^2-\alpha_n^2)\tanh(\gamma_{2,n}h)]/\Delta$$

and

$$\Delta = [\gamma_{1,n}\tanh(\gamma_{1,n}d) + \frac{\varepsilon_1}{\varepsilon_2}\gamma_{2,n}\tanh(\gamma_{2,n}h)][(\gamma_{1,n}\coth(\gamma_{2,n}d) \quad (3.37.d)$$
$$+ \frac{\mu_1}{\mu_2}\gamma_{2,n}\coth(\gamma_{2,n}h)]$$

Note that since factor $e^{-j\beta z}$ does not play a role in the following discussions, it has been omitted from Eqs. (3.36). In other words

$$\tilde{E}_{x,1} = \tilde{E}_{x,2} = \tilde{E}_x e^{-j\beta z}$$
$$\tilde{E}_{z,1} = \tilde{E}_{z,2} = \tilde{E}_z e^{-j\beta z} \qquad \text{at} \quad y=d \qquad (3.38)$$

The two algebraic equations in (3.36) are the final equations of the problem. As will be shown in the next section, they represent the Fourier transforms of two coupled integral equations in the space domain. These integral equations are solved using the remaining conditions given by Eqs. (3.28). It is worth noting that except for the factors $\pm j\omega/k_2^2$, the expressions obtained for $G_{11}(\alpha_n,\beta)$, $G_{12}(\alpha_n,\beta)$ etc., Eqs. (3.37), are the same as those reported in [27]. The factors appearing in Eqs. (3.37) are due to a different representation of the boundary conditions, Eqs. (3.21), and are also due to a slightly different definition for the Fourier transform, Eq. (3.11), in this book. These factors do not affect the final solution of the problem.

3.2.4 Final Equations in the Space Domain

In order to derive the space domain representations of Eqs. (3.36), we first observe

CHAP. 3 INTRODUCTION TO THE SPECTRAL DOMAIN TECHNIQUE

that these equations represent the spectral components of E_x and E_z. Thus, analogous to Eqs. (3.3), it is possible to write:

$$E_z = \sum_{n=-\infty}^{+\infty} G_{11}(\alpha_n,\beta)\tilde{J}_x e^{-j\alpha_n x} + \sum_{n=-\infty}^{+\infty} G_{12}(\alpha_n,\beta)\tilde{J}_z e^{-j\alpha_n x} \quad (3.39.\text{a})$$

$$E_x = \sum_{n=-\infty}^{+\infty} G_{21}(\alpha_n,\beta)\tilde{J}_x e^{-j\alpha_n x} + \sum_{n=-\infty}^{+\infty} G_{22}(\alpha_n,\beta)\tilde{J}_z e^{-j\alpha_n x} \quad (3.39.\text{b})$$

We then make use of the Fourier transform definition, Eq. (3.11.a) to transform J_x and J_z, ie:

$$\tilde{J}_x = \frac{1}{2a} \int_{-a}^{+a} J_x(\eta) e^{j\alpha_n \eta} d\eta \quad (3.40.\text{a})$$

$$\tilde{J}_z = \frac{1}{2a} \int_{-a}^{+a} J_z(\eta) e^{j\alpha_n \eta} d\eta \quad (3.40.\text{b})$$

Using these equations in Eqs. (3.39) and after some straightforward mathematical manipulation, it is possible to arrive at the following integral equations:

$$E_z = \int_{-a}^{+a} K_{11}(x-\eta) J_x(\eta) d\eta + \int_{-a}^{+a} K_{12}(x-\eta) J_z(\eta) d\eta \quad (3.41.\text{a})$$

$$E_x = \int_{-a}^{+a} K_{21}(x-\eta) J_x(\eta) d\eta + \int_{-a}^{+a} K_{22}(x-\eta) J_z(\eta) d\eta \quad (3.41.\text{b})$$

where the kernels are

$$K_{11}(x-\eta) = \frac{1}{2a} \sum_{n=-\infty}^{+\infty} G_{11}(\alpha_n,\beta) e^{-j\alpha_n(x-\eta)} \quad (3.42.\text{a})$$

$$K_{12}(x-\eta) = \frac{1}{2a} \sum_{n=-\infty}^{+\infty} G_{12}(\alpha_n,\beta) e^{-j\alpha_n(x-\eta)} \quad (3.42.\text{b})$$

$$K_{21}(x-\eta) = \frac{1}{2a} \sum_{n=-\infty}^{+\infty} G_{21}(\alpha_n,\beta) e^{-j\alpha_n(x-\eta)} \quad (3.42.\text{c})$$

$$K_{22}(x-\eta) = \frac{1}{2a} \sum_{n=-\infty}^{+\infty} G_{22}(\alpha_n,\beta) e^{-j\alpha_n(x-\eta)} \qquad (3.42.d)$$

Once again, it is worth emphasizing that E_x and E_z show the variation of the tangential electric field components with x at the air-dielectric boundary; ie: $E_x = E_x(x)$ and $E_z = E_z(x)$.

Eqs. (3.41.a) and (3.41.b) are (when applied to the boundary conditions (3.28)) a pair of coupled Fredholm's integral equations whose Fourier transforms are Eqs. (3.36.a) and (3.36.b). Since the kernels of these equations are functions of x-η, they can fall into the category of convolution integral equations [28]. This latter classification becomes more meaningful when $a \rightarrow \infty$. For those readers who are not familiar with the concept of integral equations, a brief account of preliminaries on the subject is given in Appendix II. However, in short it can be said that in an integral equation, the unknown function and a known function (kernel) appear as the integrand of an integral expression. For example, in Eq. (3.41), $J_x(\eta)$ and $J_z(\eta)$ are the functions to be determined and $K_{11}(x-\eta)$, $K_{12}(x-\eta)$ etc. are the known functions. Detailed theory of integral equations can be found in various sources including in [28,29].

A careful examination of the integral equations (3.41.a) and (3.41.b) reveals that $K_{11}(x-\eta)$, $K_{12}(x-\eta)$ etc. are the Green's functions associated with E_x and E_z. In other words if the current source

$$\mathbf{J} = (\mathbf{i} + \mathbf{k})\delta(x-x_0) \qquad (3.43)$$

is considered at the air-dielectric interface, then the resulting tangential electric field at the interface (E_x and E_z) will be given by

$$E_z = K_{11}(x-x_0) + K_{12}(x-x_0) \qquad (3.44.a)$$

$$E_x = K_{21}(x-x_0) + K_{22}(x-x_0) \qquad (3.44.b)$$

In view of the above discussion, functions $G_{11}(\alpha_n,\beta)$, $G_{12}(\alpha_n,\beta)$ etc., Eqs. (3.37), are usually referred to as the spectral domain Green's functions of microstrip-type problems. In a more compact form, Eqs. (3.36) can be shown as follows:

$$[G(\alpha_n,\beta)][\tilde{J}] = [\tilde{E}] \qquad (3.45)$$

where

CHAP. 3 INTRODUCTION TO THE SPECTRAL DOMAIN TECHNIQUE

$$[G(\alpha_n,\beta)] = \begin{bmatrix} G_{11}(\alpha_n,\beta) & G_{12}(\alpha_n,\beta) \\ G_{21}(\alpha_n,\beta) & G_{22}(\alpha_n,\beta) \end{bmatrix} \quad (3.46)$$

$$[\tilde{J}] = \begin{bmatrix} \tilde{J}_x \\ \tilde{J}_z \end{bmatrix} \quad (3.47)$$

$$[\tilde{E}] = \begin{bmatrix} \tilde{E}_z \\ \tilde{E}_x \end{bmatrix} \quad (3.48)$$

From Eq. (3.45), it is clear that $[G(\alpha_n,\beta)]$ is a dyadic Green's function [12]. Also from the same equation, it can be deduced that it is possible to construct integral equations similar to those presented in Eqs. (3.41) where the unknown functions can be an arbitrary pair from E_x, E_z, J_x and J_z. For example, take the tangential electric field components as the unknown functions. In this case, the associated integral equations in the Fourier domain are:

$$[H(\alpha_n,\beta)] [\tilde{E}] = [\tilde{J}] \quad (3.49)$$

where

$$[H(\alpha_n,\beta)] = [G(\alpha_n,\beta)]^{-1} \quad (3.50)$$

For some of its advantages, Eq. (3.49) is largely employed in the analysis of planar structures with narrow slots [30].

In fact, with a difference in the choice of the unknown functions, there are many integral equation solutions of planar structures with the final equations similar to Eqs. (3.41) (or similar to Eqs. (3.36) in the spectral domain). For instance, Minor et al [22] in the analysis of shielded microstrip line with ferrite substrate, and Yamashita et al [20] in the treatment of a more general class of planar structures, express the final integral equations in terms of E_x and J_z. Examples of using J_x and J_z or E_x and E_z in the analyses of some planar structures can be found in the works of many investigators including [1,2,18,23,27,30]. In one particular development associated with the derivation of a closed-form expression for the characteristic impedance of the microstrip line, Hashimoto [31] sets up the integral equations in terms of charge and current distributions on the strip.

CHAP. 3 INTRODUCTION TO THE SPECTRAL DOMAIN TECHNIQUE   38

Having considered the many options available to express the final integral equations in the space or in the Fourier domain, a question immediately arises as to what combination of the unknown functions (ie: J_x and J_z, E_x and E_z etc.) is the most suitable choice for the spectral domain solution of a particular problem. The answer to this, in the form of a set of criteria limiting the number of choices, will be given later in the next chapter.

3.3 Methods of Solution of the Final Equations

There are several techniques applicable for the solution of the integral equations (3.41.a) and (3.41.b). All these techniques make use of the constraints imposed on the current and field distributions at the air-dielectric interface, Eqs. (3.28). Prior to describing the method of solution associated with the spectral domain technique, it is instructive to examine the point matching method of solution. In any method employed, the aim is to convert the integral equations into a homogeneous system of linear equations. When the determinant of this system is equated to zero, the characteristic equation (the equation yielding the propagation constants of modes) of the microstrip line is obtained.

3.3.1 Point Matching (Point Collocation) Method

In this method, we begin with recalling the condition that the tangential electric field vanishes over the strip; see Eqs. (3.28.a) and (3.28.b). When this is imposed on the integral equations (3.41.a) and (3.41.b) or in fact on the equivalent equations, Eqs. (3.39), the result is the following equations:

$$\sum_{n=-\infty}^{+\infty} G_{11}(\alpha_n,\beta)\tilde{J}_x e^{-j\alpha_n x} + \sum_{n=-\infty}^{+\infty} G_{12}(\alpha_n,\beta)\tilde{J}_z e^{-j\alpha_n x} = 0 \qquad (3.51.a)$$

$$0 < |x| < w$$

$$\sum_{n=-\infty}^{+\infty} G_{21}(\alpha_n,\beta)\tilde{J}_x e^{-j\alpha_n x} + \sum_{n=-\infty}^{+\infty} G_{22}(\alpha_n,\beta)\tilde{J}_z e^{-j\alpha_n x} = 0 \qquad (3.51.b)$$

Now assume that the components of the strip current can be expanded in terms of

CHAP. 3 INTRODUCTION TO THE SPECTRAL DOMAIN TECHNIQUE

sets of basis functions:

$$J_x(x) = \sum_{p=1}^{\infty} a_p J_{x,p}(x) \qquad (3.52.a)$$

$$J_z(x) = \sum_{q=1}^{\infty} b_q J_{z,q}(x) \qquad (3.52.b)$$

where the basis functions $J_{x,p}(x)$ and $J_{z,q}(x)$ are chosen so that they are non-zero over the strip only; see Eq. (3.28.c) for the condition on the current distribution. Substituting the Fourier transforms of the above functions in Eqs. (3.51) leads to the following equations:

$$\sum_p a_p \sum_n G_{11}(\alpha_n, \beta) \tilde{J}_{x,p} e^{-j\alpha_n x} + \sum_q b_q \sum_n G_{12}(\alpha_n, \beta) \tilde{J}_{z,q} e^{-j\alpha_n x} = 0 \qquad (3.53.a)$$

$$0 < |x| < w$$

$$\sum_p a_p \sum_n G_{21}(\alpha_n, \beta) \tilde{J}_{x,p} e^{-j\alpha_n x} + \sum_q b_q \sum_n G_{22}(\alpha_n, \beta) \tilde{J}_{z,q} e^{-j\alpha_n x} = 0 \qquad (3.53.b)$$

It is not difficult to show that by giving values to x from the interval $[0, w]$, the above expressions can be turned into a set of linear homogeneous equations in unknowns a_p and b_q. The essence of this procedure is in matching E_x and E_z at discrete points over the strip boundary where these components are zero. It is clear that the number of equations obtained is equal to the number of points, which incidentally governs the total number of unknowns $(P+Q)$ which can be determined. Note that the infinite series in Eqs. (3.52.a) and (3.52.b) are assumed to be truncated to P and Q terms respectively.

Since the resulting system of equations is homogeneous, the trivial solutions ($a_p = 0$, $p=1, ..., P$ and $b_q = 0$, $q=1, ..., Q$) are not acceptable. To seek the nontrivial solutions, the determinant of coefficients must be forced to vanish. As a result, the characteristic equation is obtained whose roots are the phase constants (β) associated with discrete modes of the microstrip line. Having computed β, by a back substitution, a_p and b_q, $J_x(x)$ and $J_z(x)$ and the field components can be fully determined.

In connection with the analysis of some planar structures using similar formulation as that given for the microstrip line here, several works are reported in the literature that employ the point matching technique to handle the solution of the integral equations.

Krage et al [32] compute the phase constants of the microstrip line and the coupled microstrip line by expanding $J_x(x)$ and $J_z(x)$ in terms of trigonometric functions and Legendre polynomials respectively. For the even mode of the microstrip line, these expansions are given as follows:

$$J_x(x) = \sum_p a_p \sin \frac{p\pi x}{w} \qquad (3.54.a)$$

$$J_z(x) = \sum_q b_q P_{2(q-1)}\left(\frac{x}{w}\right) \qquad (3.54.b)$$

where $P_{2(q-1)}$ indicates the Legendre polynomials of even degree. In Krage et al [32] the system of equations is formed by matching the final equations, which are given directly as Eqs. (3.53), at equal distance points over the strip.

For the analysis of the open microstrip line, Denlinger [18] also employs the point matching technique to solve the integral equations. But he expresses $J_z(x)$ in terms of one function only,

$$J_z(x) = J_{z0}\left(1 + \left|\frac{x}{w}\right|^3\right) \qquad (3.55)$$

while neglecting altogether the contribution of the x-component of the strip current in the solution. As a result, the system of equations is reduced to one equation given by

$$J_{z0}\int_{-\infty}^{+\infty} G_{12}(\alpha,\beta)f(\alpha)d\alpha = 0 \qquad (3.56)$$

where $f(\alpha)$ is the Fourier transform of the expression in the bracket in Eq. (3.55). Note that to arrive at Eq. (3.56), Eqs. (3.53) needs to be matched at $x=0$. Also note that since the microstrip line is not shielded, α_n in Eqs. (3.53) is replaced by α, the sums are replaced by integrals and furthermore, the spectral domain Green's functions in these equations must be modified. In Denlinger's solution of the microstrip line, the accuracy of the results depends on the choice of the expression for $J_z(x)$. For this reason the given current, Eq. (3.55), is so selected to closely approximate the physical form of the current distribution over the strip. The approximate solution presented by Denlinger is not capable of predicting higher order modes and fails to be accurate at high frequencies where the contribution of $J_x(x)$ is not negligible. But it is computationally efficient.

Although not explicitly stated in their paper, Yamashita et al [20] also use the point matching method to accurately solve the integral equations obtained for a wide class of planar structures. In their work, the integral equations are given in terms of unknown functions $E_x(x)$ and $J_z(x)$ and the basis functions are chosen to be pulse functions. Each pulse is associated with a subdivision of non-uniformly discretised intervals. Intervals represent strips or slots at the air-dielectric interface and are subdivided non-uniformly, Fig. 3.4.a, so as to enhance the rate of the convergence of the solution. The reason behind using non-uniform discretization is related to the behaviour of the field near sharp edges. This effect is already discussed in section 2.4. Thus fine discretization near the strip edge would improve the rate of convergence. For the microstrip line, rates of convergence associated with the uniform and non-uniform discretizations are shown in Fig. 3.4.b [20].

In the point matching technique of solution of integral equations resulting from the mathematical modelling of an electromagnetic problem, the accuracy of the solution may be increased by increasing the number of points matched. This, however, depends on the nature of the problem and is not always guaranteed [33-35]. On the other hand, if the number of points matched is small, the point matching procedure may also fail to produce a result. For instance, the problem reported in the work of Krage et al [32], where the determinant of coefficients occasionally fails to give the phase constant, may have stemmed from this deficiency of the point matching technique.

The next technique, to be explained in the following section, handles the solution of the obtained integral equations in a more elegant manner. It utilizes the Galerkin technique together with Parseval's identity to generate a set of homogeneous linear equations. This combination has now become a specific feature in the spectral domain solution of planar structures.

It should be pointed out, however, that Galerkin's technique and the point matching (point collocation) method have the same philosophical base in the numerical analysis and in fact, they are both categorised as the moment method [33, 36-38]. Unlike the point matching method, the Galerkin technique can be applied in the Fourier and the space domains in order to convert Eqs. (3.36) and Eqs. (3.41) to the same set of homogeneous linear equations. We recall that Eqs. (3.36) are the Fourier transforms of the integral equations (3.41). Both the Fourier and the space domain approaches will be discussed here. However, prior to this, a brief summary of the moment method seems necessary for those readers unfamiliar with this subject.

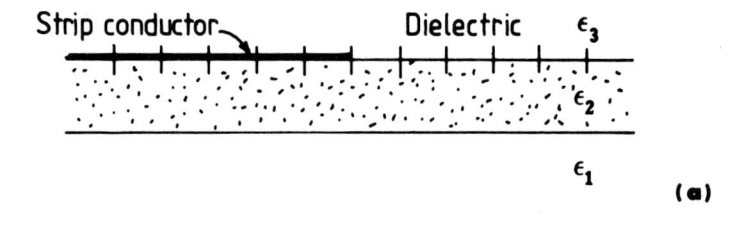

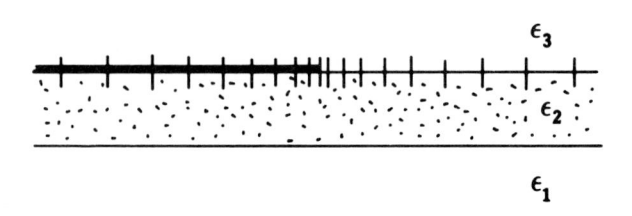

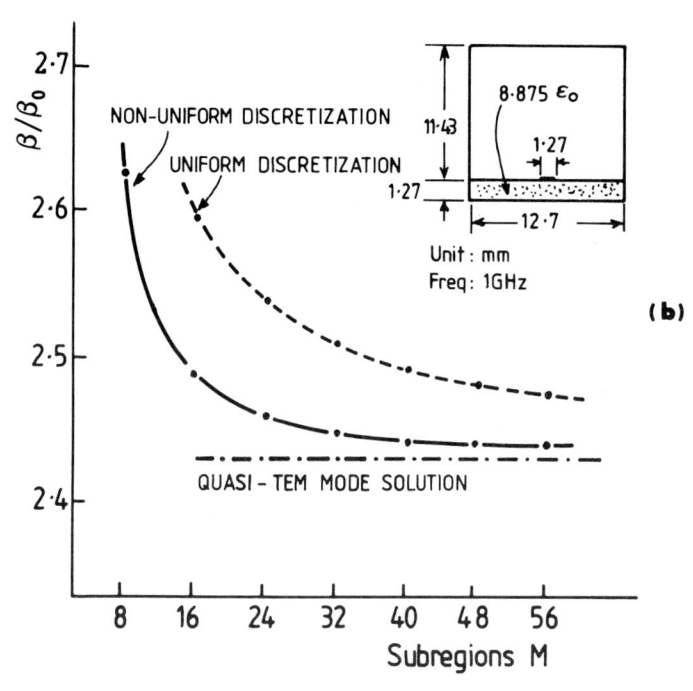

Fig. 3.4 (a) Uniform and nonuniform discretization of integration domains (strips and slots), (b) dependence of the convergence of normalized phase constant on the discretization. (From Yamashita and Atsuki [20], Copyright © 1976 IEEE, reproduced with permission.)

CHAP. 3 INTRODUCTION TO THE SPECTRAL DOMAIN TECHNIQUE

3.3.2 The Method of Moments

In the method of moments [37], which is widely used in the approximate solution of many differential and integral equations, the unknown function f is expanded in terms of a set of known functions f_p;

$$f = \sum_{p=1}^{\infty} a_p f_p \qquad (3.57)$$

The known functions f_p are commonly called the basis or trial functions and must be members of a complete set. Completeness of a set of functions can be determined by applying the Cauchy Criterion [39]. It is clear that in Eq. (3.57), a_p is unknown.

Now assume that the integral or the differential equation that we are concerned with is given as:

$$Lf = g \qquad (3.58)$$

where L represents an integral or a differential or an integro-differential operator; eg:

(i) $\quad L = \nabla^2 = \dfrac{\partial^2}{\partial x^2} + \dfrac{\partial^2}{\partial y^2} + \dfrac{\partial^2}{\partial z^2}$,

(ii) $\quad L = \iiint dx\,dy\,dz\, K(x,y,z)$

Assuming that L is a linear operator, by substituting Eq. (3.57) in Eq. (3.58), we obtain the following equation:

$$\sum_{p=1}^{\infty} a_p L f_p = g \qquad (3.59)$$

By comparing the above equation with Eqs. (3.53), it is readily concluded that the unknown coefficients a_p can be found using the point matching technique introduced in section 3.3.1. A different approach which embraces the point matching method as a special case can be developed for determining a_p. For this development, however, we need to introduce the notion of the inner product of functions first.

The concept of inner products of two functions is very similar to the concept of the inner product of two vectors. Assume a set of functions $X,Y,Z,...$ and let the inner product of two such functions, say X and Y, be denoted by $<X, Y>$. The inner product of two functions can be defined in several ways. However, no matter what definition we take for the inner product, it must satisfy the following conditions:

CHAP. 3 INTRODUCTION TO THE SPECTRAL DOMAIN TECHNIQUE

$$\langle \xi X + \zeta Y, Z \rangle = \xi \langle X, Z \rangle + \zeta \langle Y, Z \rangle \tag{3.60.a}$$

$$\langle X, Y \rangle = \langle Y, X \rangle^* \tag{3.60.b}$$

$$\langle X, X \rangle = |X|^2 \tag{3.60.c}$$

where (*) specifies the complex conjugate and ξ, ζ are scalar quantities. For further reading on the inner products, the reader may refer to [39]. Analogous to vectors, two functions are said to be orthogonal when their inner product is zero. The concept of orthogonality of two functions can be exploited in order to convert Eq. (3.59) into a set of linear equations. To this end, we argue that if the function f is the exact solution of Eq. (3.58), then the expression

$$\sum_{p=1}^{\infty} a_p L f_p - g \tag{3.61}$$

must be orthogonal to any member of a set of functions called testing functions t_i. In other words:

$$\langle t_i, \sum_{p=1}^{\infty} a_p L f_p - g \rangle = 0 \qquad i=1,2,\ldots \tag{3.62}$$

Using properties of inner products given by Eqs. (3.60), the above equation can be written as follows:

$$\sum_{p=1}^{\infty} a_p^* \langle t_i, L f_p \rangle = \langle t_i, g \rangle \qquad i=1,2,\ldots \tag{3.63}$$

From the above equation, it is clear that by taking the inner product of both sides of Eq. (3.59) with the testing functions, a linear system of equations results, Eq. (3.63), from which the unknown parameters a_p can be determined. The way that the basis and testing functions are defined varies from one problem to another. A judicious choice of these functions (a) enhances the accuracy of the solution, (b) facilitates the evaluation of elements of the matrix (note that Eq. (3.63) can be written in a matrix form), (c) reduces the size of the matrix for a required accuracy of solution and (d) improves the matrix condition [36,37]. Hence, the overall result is the use of less computer memory and time for approximating the function f.

As can be gathered from the above discussion, there should exist an infinite number of basis functions and testing functions which can be employed in the method of moments for approximating a function. Considering this fact, two specializations of this technique are as follows [36-38].

CHAP. 3 INTRODUCTION TO THE SPECTRAL DOMAIN TECHNIQUE

(i) *Galerkin's technique*

In the Galerkin technique, the testing functions are the same as the basis functions ($t_i = f_i$). In the solutions of many problems, this choice offers many advantages including some of those referred to above.

(ii) *Point matching (collocation) technique*

When testing functions are delta functions; eg: $t_i = \delta(x-x_i)$, the moment method is widely known as the point matching method.

In introducing the method of moments in this section, an attempt has been made to omit the mathematical details and to just give an overview of the technique. The material covered, however, is believed to be sufficient for understanding of many numerical solutions of electromagnetic problems, based on the method of moments.

We now turn to the solution of the integral equations (3.41) using Galerkin's technique in the space domain initially and then employing it in the spectral domain in order to solve the same equations.

3.3.2.1 Galerkin Technique in the Space Domain

We begin the Galerkin method of solution of Eqs. (3.41) by expanding $J_x(x)$ and $J_z(x)$ in terms of a complete set of basis functions. These expansions are already given in Eqs. (3.52), so using them in Eqs. (3.41) leads to the following equations:

$$E_z = \sum_p a_p \int_{-a}^{+a} K_{11}(x-\eta) J_{x,p}(\eta) d\eta + \sum_q b_q \int_{-a}^{+a} K_{12}(x-\eta) J_{z,q}(\eta) d\eta \quad (3.64.a)$$

$$E_x = \sum_p a_p \int_{-a}^{+a} K_{21}(x-\eta) J_{x,p}(\eta) d\eta + \sum_q b_q \int_{-a}^{+a} K_{22}(x-\eta) J_{z,q}(\eta) d\eta \quad (3.64.b)$$

In order to be able to apply the method of moments to the above equations, we need to define an appropriate inner product. Since we are particularly interested in the Galerkin technique, a suitable inner product is as follows:

$$<X,Y> = \frac{1}{2a}\int_{-a}^{+a} X(x)Y^*(x)dx \qquad (3.65)$$

One may check that the above expression satisfies conditions (3.60).

The final step in the Galerkin technique is to take the inner products of Eqs. (3.64) with the basis functions. In this process, we choose to take the inner product of Eq. (3.64.a) with $J_{z,q}(x)$ and Eq. (3.64.b) with $J_{x,p}(x)$. The result is a set of homogeneous linear equations given as follows:

$$\sum_{p=1}^{P} C_{\acute{q},p}^{1,1}(\beta) a_p + \sum_{q=1}^{Q} C_{\acute{q},q}^{1,2}(\beta) b_q = 0 \qquad \acute{q} = 1,\ldots,Q \qquad (3.66.a)$$

$$\sum_{p=1}^{P} C_{\acute{p},p}^{2,1}(\beta) a_p + \sum_{q=1}^{Q} C_{\acute{p},q}^{2,2}(\beta) b_q = 0 \qquad \acute{p} = 1,\ldots,P \qquad (3.66.b)$$

where

$$C_{\acute{q},p}^{1,1}(\beta) = <\int_{-a}^{+a} K_{11}(x-\eta)J_{x,p}(\eta)d\eta, J_{z,\acute{q}}> = \sum_{n} G_{11}(\alpha_n,\beta)\tilde{J}_{z,\acute{q}}^* \tilde{J}_{x,p} \qquad (3.67.a)$$

$$C_{\acute{q},q}^{1,2}(\beta) = <\int_{-a}^{+a} K_{12}(x-\eta)J_{z,q}(\eta)d\eta, J_{z,\acute{q}}> = \sum_{n} G_{12}(\alpha_n,\beta)\tilde{J}_{z,\acute{q}}^* \tilde{J}_{z,q} \qquad (3.67.b)$$

$$C_{\acute{p},p}^{2,1}(\beta) = <\int_{-a}^{+a} K_{21}(x-\eta)J_{x,p}(\eta)d\eta, J_{x,\acute{p}}> = \sum_{n} G_{21}(\alpha_n,\beta)\tilde{J}_{x,\acute{p}}^* \tilde{J}_{x,p} \qquad (3.67.c)$$

$$C_{\acute{p},q}^{2,2}(\beta) = <\int_{-a}^{+a} K_{22}(x-\eta)J_{z,q}(\eta)d\eta, J_{x,\acute{p}}> = \sum_{n} G_{22}(\alpha_n,\beta)\tilde{J}_{x,\acute{p}}^* \tilde{J}_{z,q} \qquad (3.67.d)$$

Note that the inner products in Eqs. (3.67) are evaluated using Parseval's identity [16, Appendix I]:

$$\frac{1}{2a}\int_{-a}^{+a} X(x)Y^*(x)dx = \sum_{n} \tilde{X}\tilde{Y}^* \qquad (3.68)$$

Also note that because of conditions (3.28) at the interface, we have

CHAP. 3 INTRODUCTION TO THE SPECTRAL DOMAIN TECHNIQUE

$$<E_z, J_{z,q'}> = 0 \quad (3.69.a)$$

$$<E_x, J_{x,p'}> = 0 \quad (3.69.b)$$

The system (3.66) is the desired set of equations. In compact form it can be written as follows:

$$[C(\beta)] \begin{bmatrix} a \\ \hline b \end{bmatrix} = [0] \quad (3.70)$$

where $[C(\beta)]$ is a matrix of order $P+Q$ whose elements appear in Eqs. (3.67). Due to computer limitation in handling infinite size matrices, $P+Q$ has a finite value. The column matrix following the coefficient matrix $[C(\beta)]$ contains a_p and b_q in the following form:

$$\begin{bmatrix} a \\ \hline b \end{bmatrix} = \begin{bmatrix} a_1 \\ a_2 \\ \vdots \\ \hline b_1 \\ b_2 \\ \vdots \end{bmatrix} \quad (3.71)$$

Considering Eq. (3.70), it is clear that its nontrivial solutions are given when

$$\det [C(\beta)] = 0 \quad (3.72)$$

from which the phase constant β can be computed. Using Eq. (3.70), with a similar strategy as that mentioned in section 3.3.1, initially a_p and b_q and subsequently by employing Eqs. (3.52), (3.26), (3.25) and (3.24), all the field components associated with a phase constant can be determined.

3.3.2.2 Galerkin Technique in the Spectral Domain

The starting point in the application of Galerkin's method in the spectral domain for the solution of the integral equations (3.41) is Eqs. (3.36). We recall that the latter equations are the Fourier transforms of the former ones.

We again use expansions (3.52) and substitute them in Eqs. (3.36). The result is the following set of equations in the spectral domain:

$$\tilde{E}_z = \sum_p a_p G_{11}(\alpha_n,\beta)\tilde{J}_{x,p} + \sum_q b_q G_{12}(\alpha_n,\beta)\tilde{J}_{z,q} \quad (3.73.a)$$

$$\tilde{E}_x = \sum_p a_p G_{21}(\alpha_n,\beta)\tilde{J}_{x,p} + \sum_q b_q G_{22}(\alpha_n,\beta)\tilde{J}_{z,q} \quad (3.73.b)$$

Let the inner product required for the application of the Galerkin technique in the Fourier domain with discrete spectrum be defined by

$$<\tilde{X},\tilde{Y}> = \sum_n \tilde{X}\tilde{Y}^* \quad (3.74)$$

For the Fourier transform with continuous spectrum, the above definition can be generalised as follows:

$$<\tilde{X},\tilde{Y}> = \int_{-\infty}^{+\infty} \tilde{X}\tilde{Y}^* d\alpha \quad (3.75)$$

This inner product is suitable for the spectral domain solution of planar structures with side walls moved to infinity $(a \to \infty)$; see Fig. 3.2.a and Fig. 3.2.b.

With the definition given in Eq. (3.74), we are now in a position to apply the Galerkin technique to Eqs. (3.73) in the spectral domain. In order to form a set of equations as given in Eqs. (3.66), we take the inner product of Eq. (3.73.a) with the Fourier transform of J_z and Eq. (3.73.b) with the Fourier transform of J_x. The result is as follows:

$$\sum_{p=1}^{P} C_{q,p}^{1,1}(\beta) a_p + \sum_{q=1}^{Q} C_{q,q}^{1,2}(\beta) b_q = \sum_n \tilde{J}_{z,\dot{q}}^* \tilde{E}_z \quad \dot{q} = 1,...,Q \quad (3.76.a)$$

$$\sum_{p=1}^{P} C_{\dot{p},p}^{2,1}(\beta) a_p + \sum_{q=1}^{Q} C_{\dot{p},q}^{2,2}(\beta) b_q = \sum_n \tilde{J}_{x,\dot{p}}^* \tilde{E}_x \quad \dot{p} = 1,...,P \quad (3.76.b)$$

Using Parseval's identity, Eq. (3.68), and the boundary conditions (3.28), it is not difficult to show that the summations on the right-hand side of Eqs. (3.76) are zero, eg:

$$\sum_n \tilde{J}_{z,\dot{q}}^* \tilde{E}_z = \frac{1}{2a} \int_{-a}^{+a} J_{z,\dot{q}}^*(x) E_z dx = 0 \quad (3.77)$$

Although the application of the Galerkin technique in the space or in the Fourier domain effectively leads to the same set of Eqs. (3.66), it seems the latter procedure is more convenient and straight-forward. Also, as far as the name "spectral domain technique" is concerned, the Fourier domain application of the Galerkin technique is more meaningful.

At this stage we can claim that the spectral domain formulation of the microstrip line is complete. Various points as regards the types of basis functions, approximate solutions etc. which are more relevant to the numerical side of the spectral domain technique will be elaborated in a general sense in the following chapters.

References

1. Jansen R.H., "The spectral domain approach for microwave integrated circuits", *IEEE Trans. Microwave Theory Tech.*, **MTT-33**, pp.1043-1056, 1985.

2. Jansen R.H., "Unified user-oriented computation of shielded, covered and open planar microwave and millimeter-wave transmission-line characteristics", *IEE J. Microwave, Opt. Acoust.*, **Vol. MOA-3**, pp.14-22, 1979.

3. Alexopoulos N.G., "Integrated-circuit structures on anisotropic substrates", *IEEE Trans. Microwave Theory Tech.*, **MTT-33**, pp.847-881, 1985.

4. Nakatani A. and Alexopoulos N.G., "A generalised algorithm for the modelling of the dispersive characteristics of microstrip, inverted microstrip, stripline, slotline, finline, and coplanar waveguide circuits on anisotropic substrate", *IEEE* **MTT-S Digest**, pp.457-459, 1985.

5. Tsalamengas J.L., Uzunoglu N.K. and Alexopoulos N.G., "Propagation characteristics of a microstrip line printed on a general anisotropic substrate", *IEEE Trans. Microwave Theory Tech.*, **MTT-33**, pp.941-945, 1985.

6. Marques R. and Horno M., "On the spectral dyadic Green's function for stratified linear media - Application to multilayer MIC lines with anisotropic dielectrics", *IEE Proceedings,* **Vol. 134**, Pt. H, No.3, pp.241-248, 1987.

7. Manzur J. and Grabowski K., "Spectral domain analysis of multilayered transmission lines with anisotropic media", *URSI Symp. on EMW*, pp.233 C/1-223 C/4, 1980.

8. Nakatani A. and Alexopoulos N.G., "Towards a generalised algorithm for the modelling of the dispersive properties of integrated circuit structures on anisotropic substrates", *IEEE Trans. Microwave Theory and Tech.*, **MTT-33**, pp.1436-1441, 1985.

9. Hayashi Y. and Kitazawa T., "Analysis of microstrip transmission line on a sapphire substrate", *J. Inst. Electron. Commun. Eng. Jap.*, **Vol. 62-B**, pp.596-602, 1979.

10. Mirshekar-Syahkal D., "An accurate determination of dielectric loss effect in MMICs including microstrip and coupled microstrip lines", *IEEE Trans. Microwave Theory Tech.*, **MTT-31**, pp.950-954, 1983.

11. Mu T., Ogawa H. and Itoh T., "Characteristics of multiconductor, asymmetric, slow-wave microstrip transmission lines", *ibid*, **MTT-34**, pp.1471-1477, 1986.

12. Collin R.E., *Field Theory of Guided Waves*, McGraw Hill, New York, 1966.

13. Mittra R. and Itoh T., "Analysis of microstrip transmission lines" in : *Advances in Microwaves*, Vol. 8, Academic Press, New York, 1974.

14. Lighthill M.J., *An Introduction to Fourier Analysis and Generalized Functions*, Cambridge University Press, Cambridge, 1958.

15. Lanczos C., *Discourse on Fourier Series*, Oliver and Boyd, Edinburgh & London, 1966.

16. Sneddon I.N., *Fourier Transforms*, McGraw-Hill, New York, 1951.

17. Zysman G.I. and Varon D., "Wave propagation in microstrip transmission lines", Int. Microwave Symp. Dig. (Dallas, U.S.A.), pp.3-9, 1969.

18. Denlinger E.J., "A frequency dependent solution for microstrip transmission lines", *IEEE Trans. Microwave Theory Tech.*, **MTT-19**, pp.30-39, 1971.

19. Mittra R. and Itoh T., "A new technique for analysis of dispersion characteristics of microstrip lines", *ibid*, **MTT-19**, pp.47-56, 1971.

20. Yamashita E. and Atsuki K., "Analysis of microstrip-like transmission lines by nonuniform discretization of integral equations", *ibid*, **MTT-24**, pp.195-200, 1976.

21. Minor J.C. and Bolle D.M., "Propagation in shielded microslot on ferrite substrate", *Electron. Lett.*, **Vol. 7**, pp.502-504, 1971.

22. Minor J.C. and Bolle D.M., "Modes in the shielded microstrip on a ferrite substrate transversely magnetized in the plane of the substrate", *IEEE Trans. Microwave Theory Tech.*, **MTT-19**, pp.570-577, 1971.

23. Farrar A. and Adams A.T., "Computation of propagation constants for the fundamental and higher order modes in microstrip", *ibid*, **MTT-24**, pp.456-460, 1976.

24. Ganguly A.K. and Speilman B.E., "Dispersion characteristics for arbitrarily configured transmission media", *ibid*, **MTT-25**, pp.1138-1141, 1977.

25. Omar A.S. and Schunemann K., "Formulation of the singular integral equation for general planar transmission lines", *IEEE* **MTT-S Digest**, pp.135-138, 1985.

26. Dekleva J. and Role V., "Accurate numerical solution of coupled integral equations for microstrip transmission lines", *IEE Proceedings*, **Vol. 134**,

Pt. H, No. 2, pp.163-168, 1987.

27. Itoh T. and Mittra R., "A technique for computing dispersion characteristics of shielded microstrip lines", *IEEE Trans. Microwave Theory Tech.*, **MTT-22**, pp. 396-397, 1974.

28. Wyld H.W., *Mathematical Methods for Physics*, The Benjamin/Cummings Publishing Co., Menlo Park, California, 1976.

29. Arsenin, V.Ya., *Basic Equations and Special Functions of Mathematical Physics*, London Iliffe Books Ltd., 1968.

30. Mirshekar-Syahkal D. and Davies J.B., "Accurate solution of microstrip and coplanar structures for dispersion and for dielectric and conductor losses", *IEEE Trans. Microwave Theory Tech.*, **MTT-27**, pp.694-699, 1979.

31. Hashimoto M., "A rigorous solution for dispersive microstrip", *ibid,* **MTT-33**, pp.1131-1137, 1985.

32. Krage M.K. and Haddad G.I., "Frequency dependent characteristics of microstrip transmission lines", *ibid,* **MTT-20**, pp.678-688, 1972.

33. Becker M., *The Principles and Applications of Variational Methods*, MIT Press, Cambridge, Mass., 1964.

34. Isaacson E. and Keller H.B., *Analysis of Numerical Methods*, John Wiley & Sons, New York, 1966.

35. Davies J.B. and Nagenthiram P., "Irregular fields, non-convex shapes and the point matching method for hollow waveguides", *Electron. Lett.*, **Vol. 7**, pp.401-404, 1971.

36. Harrington R.F., "Matrix methods for field problems", *Proceedings of the IEEE*, **Vol. 55**, No. 2, pp.136-149, 1967.

37. Harrington R.F., *Field Computation by Moment Methods*, Collier-Macmillan Ltd., London, 1968. Reprinted by R.E. Krieger Publishing Co., Florida 32950, 1982.

38. Zienkiewicz O.C. and Morgan K., *Finite Elements and Approximation*, John Wiley & Sons, New York, 1983.

39. Apostol T.M., *Mathematical Analysis*, Addison-Wesley, Reading, Mass. 1978.

Chapter 4

EFFICIENT COMPUTING AND APPROXIMATIONS IN THE SPECTRAL DOMAIN TECHNIQUE

The previous chapter was entirely devoted to introducing the spectral domain technique and its analytical formulation for the microstrip line. No special reference was made as to how the numerical computation can be performed efficiently, nor was it attempted to disclose the factors influencing the numerical efficiency. This chapter is mainly concerned with these aspects in the spectral domain technique. The choice of basis functions and its relevance to the enhancement of the computing efficiency are explained initially. Various examples of commonly used basis functions are given and their merits are compared. Then, special relations which usually exist among the elements of the final matrix are brought to the reader's attention. These relations should be exploited in order to reduce the computer time and memory. Various choices of the spectral domain Green's functions and the optimum choice for the solution of a planar structure are the topics which are discussed next. Approximate solutions and certain hints for accelerating the computation process are also included in this chapter.

Although they may be developed around particular examples, materials in this chapter should not be specifically associated with these cases. Therefore, concluding remarks are general, being applicable to the spectral domain solution of many planar structures such as the finline, coplanar waveguide etc.

4.1 Basis Functions

The behaviour of the electromagnetic field in the vicinity of sharp edges has been discussed in section 2.4. It was shown there that in the microstrip-type problems where the strips are modelled by infinitely thin perfect conductors, Fig. 2.4, the

CHAP. 4 EFFICIENT COMPUTING AND APPROXIMATIONS IN ...   54

transverse electric and magnetic field components are unbounded at the strip edge and approach infinity with $r^{-\frac{1}{2}}$. The longitudinal field components are bounded and behave as $r^{\frac{1}{2}}$ near the edge. These asymptotic solutions are in turn reflected in the components of the current distribution of the strip (see Eqs. 3.21), so that their behaviour near the edge ($x \to w$, where w is the x coordinate of the edge of the strip) can be described as

$$J_x(x) \to (x-w)^{\frac{1}{2}} \qquad (4.1.a)$$

$$J_z(x) \to (x-w)^{-\frac{1}{2}} \qquad (4.1.b)$$

The above conclusions are general and hold at the edge of an infinitely thin strip of perfect conductor sandwiched between two dielectrics. For such a combination of strip and dielectrics, the dielectric constants do not enter the given asymptotic expressions.

The significance of direct incorporation of the edge singularities in the mathematical modelling of problems involving sharp edges is already noted in section 2.4. The principal concern in an explicit satisfaction of the edge requirement is to safeguard against converging towards a non-physical solution (so called relative convergence), although other benefits including enhancement of the rate of convergence and improvement of the matrix condition, if applicable, usually follow.

In connection with the spectral domain solution of planar structures, it should thus be expedient to choose basis functions satisfying the expected behaviour of the field or current near the strip edge [1,2]. However, in contrast with some other methods of solution of problems involving sharp edges [3-5], the spectral domain technique does not seem to suffer from the relative convergence in the event that the edge condition is not considered in the basis functions. In fact, a survey of literature disclosed nothing to the contrary. Furthermore, the author's investigations with many different basis functions have not revealed a substantial change in the results of various parameters of planar structures, to infer the existence of the relative convergence. Pozar et al [6],too, seem to have observed this virtue of the spectral domain technique. However, in the evaluation of certain parameters of planar structures such as the characteristic impedances of planar transmission lines where an accurate description of the current or field near the edge is necessary, computations can be carried out very efficiently if the basis functions satisfy the edge condition [1].

As mentioned earlier in section 3.3.2 under the method of moments, in order to be able to systematically increase the solution accuracy by increasing the number of basis

CHAP. 4 EFFICIENT COMPUTING AND APPROXIMATIONS IN ...   55

functions, these functions must belong to a complete set. This is an important criterion and it must be strictly observed for the basis functions chosen for the expansion of current or field in the spectral domain solution of a planar structure.

Besides the above two criteria of the edge condition and the completeness, further constraints can be imposed on the choice of basis functions in order to enhance the rate of convergence, to save the computing time and to avoid spurious solutions. Spurious solutions may be eliminated by choosing basis functions which are twice continuously differentiable [1]. Computing time can be significantly reduced if the basis functions have closed-form Fourier transforms. Rate of convergence may be speeded up by employing basis functions whose behaviours resemble the physical distribution of the expanded field or current. Jansen [1] in the analysis of microstrip lines considers a further constraint relating $J_{x,p}(x)$ and $J_{z,q}(x)$ through an integral relation. As will be explained later, although this arrangement saves computing time, it could not be generally applicable to all spectral domain solution cases.

In the following we give examples of the basis functions used by various investigators in the analysis of planar transmission lines by the spectral domain technique. Basis functions for solutions of planar structures with more complex conductor pattern such as resonators, antennas, periodic structures etc. will be presented later in appropriate sections of the book.

Many investigators have employed a combination of sinusoidal functions and functions satisfying the edge condition in order to expand the singular and regular field or current components at the interfaces where conductors are etched. For example, in the analysis of the microstrip line, Fig. 3.3, Jansen [1] proposes the following sets of basis functions for expanding the current distribution:

$$J_{z,1}(x) = (1-X^2)^{-\frac{1}{2}}$$
$$J_{z,q}(x) = \{cos[(q-1)\pi X] - J_0[(q-1)\pi]\}(1-X^2)^{-\frac{1}{2}} \quad q=2,3, ... \quad (4.2.a)$$

$$J_{x,1}(x) = 0$$
$$J_{x,p}(x) = \int_0^X J_{z,p}(X')dX' \quad p=2,3, ... \quad (4.2.b)$$

where $X = x/w$ and J_0 denotes the zero order Bessel function of the first kind. Checking the above expansion functions against the criteria introduced earlier in this section, it is evident that they satisfy all the constraints. The only deficiency which may be associated with the above set is that the Fourier transforms of its elements contain Bessel functions which could be a disadvantage from the computing point of view. For laterally open microstrip line, Fig. 3.2.b, the Fourier transforms of the above equations are given by the following expressions [1,7]:

$$\tilde{J}_{z,1} = \frac{w}{2} J_0(\alpha_w)$$
$$\tilde{J}_{z,q} = \frac{w}{4} \{ J_0[\alpha_w - (q-1)\pi] + J_0[\alpha_w + (q-1)\pi] - 2J_0[(q-1)\pi]J_0(\alpha_w) \}$$
$$q = 2,3,\ldots$$
(4.3.a)

$$\tilde{J}_{x,1} = 0$$
$$\tilde{J}_{x,p} = \frac{\tilde{J}_{z,p}}{\alpha_w} \qquad p = 2,3,\ldots$$
(4.3.b)

where $\alpha_w = \alpha w$ and α is defined in Eq. (3.11.c). Note that in the derivation of the above equations, since the microstrip is assumed open at its two sides, Eq. (3.11.c) is used to perform the Fourier transform. Also note that the expansion functions given by Eqs. (4.2) are valid for the even modes only. For the odd modes, the associated basis functions can be found in [1]. A slightly different version of the above basis functions has also appeared in some other works of Jansen [2,8] which are worth studying.

In contrast to Jansen [1,2,8], in a different manner, Minor et al [9], in the solution of the microstrip line, combine the singular function and trigonometric functions in order to express the current and the field distributions at the air-dielectric interface. Using their method of expansion, $J_z(x)$ for the even mode of the microstrip line for instance, is given by:

$$J_z(x) = b_1(1-X^2)^{-\frac{1}{2}} + \sum_{q=2}^{Q} b_q \cos[(q-2)\pi X]$$
(4.4)

where $X = x/w$. It is clear that in the above expansion the first term takes care of the singularities associated with the edges of the strip. Compared with the expansion functions (4.2.a), the Fourier transformation of the basis functions in Eq. (4.4) involves one special function only, thus taking less computer time to be generated. Using the integral relation (4.2.b), a suitable expansion for $J_x(x)$ can be automatically obtained from Eq. (4.4).

Examples of basis functions not supporting the edge condition and yet successfully employed in the spectral domain solution of various planar structures are scattered in the literature. An interesting example of such basis functions can be found in the work of Schmidt et al [10]. In that study which is associated with the analysis of modes in the bilateral finline, Fig. 4.1, the integral equations in the Fourier domain are given in terms of the slot field components E_x and E_z; see Eq. (3.49). For solution of these equations, as proposed by Schmidt et al, one possible way of expanding E_x and E_z is to use a set of pulses defined as:

$$E_{x,p}(x) = \begin{cases} 1 & (p-1)\Delta x < |x| < p\Delta x \\ 0 & \text{otherwise} \end{cases} \quad (4.5.a)$$

$$E_{z,q}(x) = \begin{cases} 1 & (q-1)\Delta x' < x < q\Delta x' \\ -1 & -q\Delta x' < x < -(q-1)\Delta x' \\ 0 & \text{otherwise} \end{cases} \quad (4.5.b)$$

where $\Delta x = s/P$, $\Delta x' = s/Q$ and s is the half width of the slot. A pictorial representation of the above functions is given in Fig 4.2. Although the Fourier transformations of the above functions are straightforward and analytically possible, the number of pulses required in order to accurately represent the field components, particularly E_x which is singular, is large. Thus, the final matrix is large and this is a disadvantage from the computing point of view. However, due to their flexible nature, the pulse-type expansion functions have an important place in the numerical solution of integral equations [11,12]. As may be recalled, we have already discussed in section 3.3.1 the application of pulse functions in the solution of integral equations of microstrip-type structures using the point matching method. An interesting feature of expanding a function in terms of pulses is that the value of the function at any point is equal to one of the expansion coefficients. In the same work, Schmidt et al [10] also present different types of expansion functions for E_x and E_z which are very similar to those given by Jansen, Eqs. (4.2), ie:

$$E_{x,p}(x) = \cos[(p-1)\pi(\frac{x}{s}+1)][1-(\frac{x}{s})^2]^{\frac{1}{2}} \quad (4.6.a)$$

$$E_{z,q}(x) = \sin[q\pi(\frac{x}{s}+1)][1-(\frac{x}{s})^2]^{\frac{1}{2}} \quad (4.6.b)$$

The first few functions of the above sets are shown in Fig. 4.3. However, results obtained through the use of basis functions given by Eqs. 4.5 show little difference

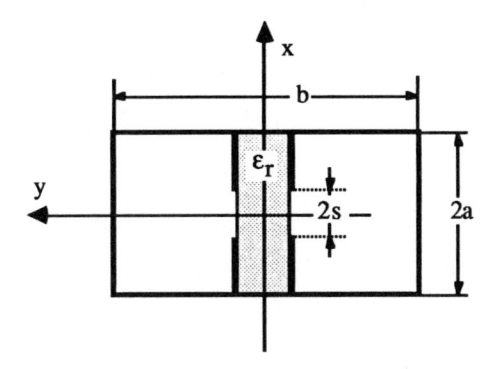

CHAP. 4 EFFICIENT COMPUTING AND APPROXIMATIONS IN ... 58

Fig. 4.1 Cross-section of a bilateral finline.

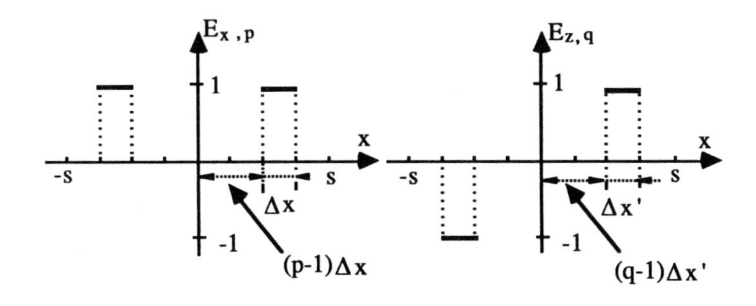

Fig. 4.2 Pulse functions as basis functions for expanding E_x and E_z. Note that $\Delta x = s/P$ and $\Delta x' = s/Q$ where P and Q are the numbers of basis functions in the expansions of $E_x(x)$ and $E_z(x)$ respectively.

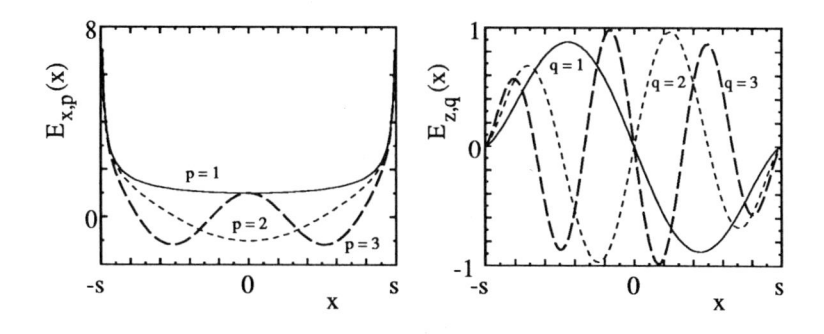

Fig. 4.3 Basis functions corresponding to (4.6.a) and (4.6.b).

CHAP. 4 EFFICIENT COMPUTING AND APPROXIMATIONS IN... 59

from those obtained by Eqs. 4.6 [10]. These results as well as many others reported in the literature are good evidence that in the spectral domain technique, explicit satisfaction of the edge condition would not be necessary. But we should not forget that its implementation in the solution undoubtedly enhances the rate of convergence and saves computing time. As an example see Fig. 4.4.

A further example of non-singular basis functions employed to expand the singular field and current can be observed in [13-15]. In those works, Legendre polynomials have been used to expand $J_z(x)$ and $E_x(x)$ and trigonometric functions employed to express the regular field or current distributions ($J_x(x)$ and $E_z(x)$). For instance for the even mode of the microstrip line, the given basis functions are as follows:

$$J_{x,p}(x) = \sin(p\pi X)$$
$$J_{z,q}(x) = P_{2(q-1)}(X) \quad 0 < |X| < 1 \quad (4.7)$$

where $X = x/w$ and $P_{2(q-1)}$ represents the Legendre polynomials of the even degree. The use of Legendre polynomials for the expansion of singular field or current may

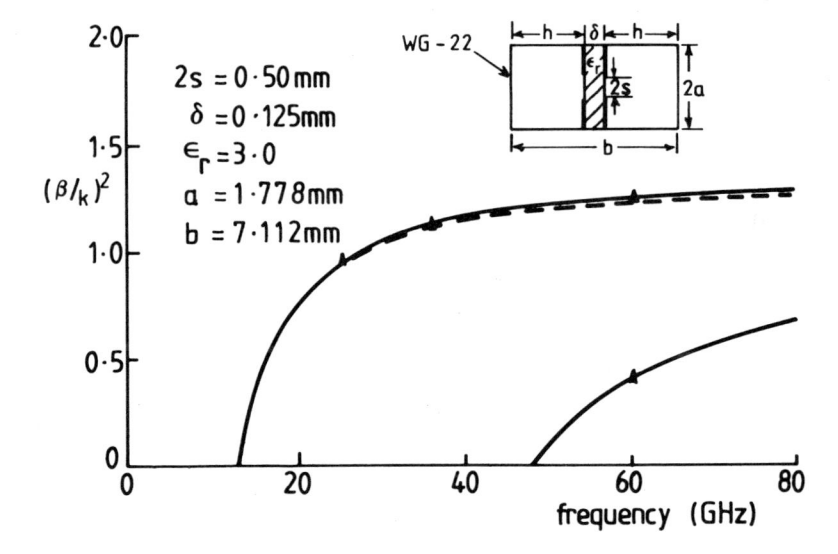

Fig. 4.4 Dispersion characteristics of the dominant and first higher order mode of a bilateral finline computed by the spectral domain technique with n=200 (n is the number of Fourier terms) for different matrix orders (P+Q) and for different basis functions: —— for P=Q=1 in (4.6), - - - for P=Q=2 in (4.6) and ▲ for P=Q=7 in (4.5). (From Schmidt and Itoh [10], Copyright © 1980 IEEE, reproduced with permssion.)

be justified on the ground that

$$\sum_{q=0}^{Q \to \infty} P_{2(q-1)}(X) \to [2(1-X^2)]^{\frac{1}{2}} \qquad (4.8)$$

where the right-hand side of the above expression indicates a singularity of the type encountered in the field or current component at the edges of the strips in the planar structures. In the case of the microstrip line, the expansion of $J_z(x)$ in terms of Legendre polynomial should be particularly advantageous. This is because, except at the edges, the behaviour of these functions over the strip, Fig. 4.5, resembles the longitudinal current distributions of discrete modes of an isolated conducting strip, Fig. 4.6. Current distributions shown in Fig. 4.6 are successfully employed by Van De Capelle et al [16] to analyse the dominant and higher order modes of open microstrip lines. Finally, so far as the Fourier transforms of Legendre polynomials are concerned, they can be obtained in closed-forms involving no special functions in the Fourier domain. In fact, since there is a recurrence relation among Legendre polynomials of three consecutive degrees, ie:

$$P_{q+1}(x) = \frac{2q+1}{q+1} x P_q(x) - \frac{q}{q+1} P_{q-1}(x) \qquad (4.9)$$

an arrangement for an efficient computer generation of the Fourier transforms of these functions is possible.

We end the discussion on the basis functions here. However, since this subject is as important as the spectral domain formulation of a particular structure itself, it will inevitably be brought up time and again in the following chapters.

4.2 Approximate Solutions

In addition to the formal solution of planar structures by the spectral domain technique in which the unknown current or field components are expressed in series of some suitable basis functions, in many practical instances it is possible to cut the solution short by employing approximate currents or fields. As we will be demonstrating in a short while, this will save us a great deal of computing time, provided that a prior knowledge of the shape of the field or the current components of concern is available.

CHAP. 4 EFFICIENT COMPUTING AND APPROXIMATIONS IN ...   61

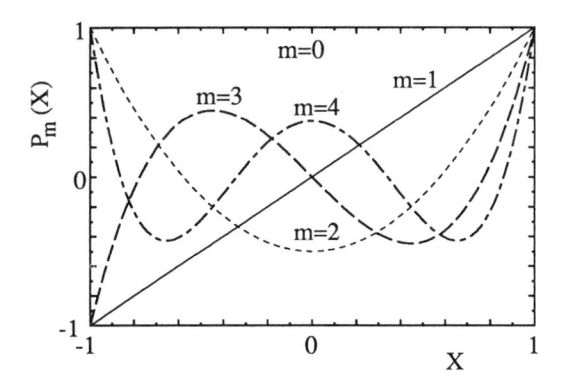

Fig. 4.5 Legendre polynomials over the interval [-1 , 1].

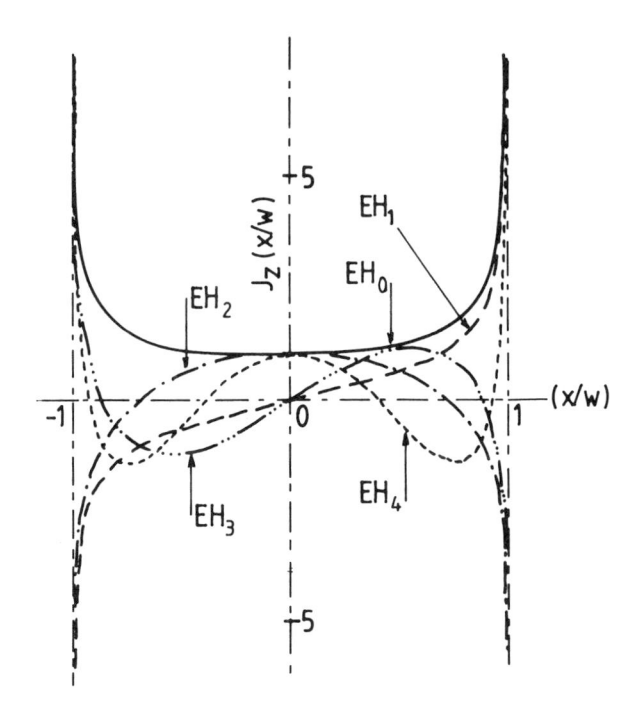

Fig. 4.6 Approximate longitudinal current distributions for the first five modes of the microstrip line. (From Van de Capelle and Luyparet [16], Electronics Letters, Vol. 9, pp.345-346, 1973, reproduced with permssion of IEE.)

CHAP. 4 EFFICIENT COMPUTING AND APPROXIMATIONS IN ...   62

Let us turn to the solution of the microstrip line in Chapter 3 and assume that for a particular mode the approximate physical distributions for $J_x(x)$ and $J_z(x)$ are available and given by the functions $j_x(x)$ and $j_z(x)$ respectively. Thus, we can write:

$$J_x(x) = a_1 j_x(x) \qquad (4.10.a)$$
$$J_z(x) = b_1 j_z(x) \qquad (4.10.b)$$

Substituting the Fourier transforms of the above expressions in Eqs. (3.66) leads to the following set of linear equations:

$$C_{1,1}^{1,1}(\beta)a_1 + C_{1,1}^{1,2}(\beta)b_1 = 0 \qquad (4.11.a)$$
$$C_{1,1}^{2,1}(\beta)a_1 + C_{1,1}^{2,2}(\beta)b_1 = 0 \qquad (4.11.b)$$

where

$$C_{1,1}^{1,1}(\beta) = \sum_n G_{11}(\alpha_n,\beta)\tilde{j}_z^* \tilde{j}_x \qquad (4.12.a)$$

$$C_{1,1}^{1,2}(\beta) = \sum_n G_{12}(\alpha_n,\beta)|\tilde{j}_z|^2 \qquad (4.12.b)$$

$$C_{1,1}^{2,1}(\beta) = \sum_n G_{21}(\alpha_n,\beta)|\tilde{j}_x|^2 \qquad (4.12.c)$$

$$C_{1,1}^{2,2}(\beta) = \sum_n G_{22}(\alpha_n,\beta)\tilde{j}_x^* \tilde{j}_z \qquad (4.12.d)$$

As explained in the previous chapter, the characteristic equation giving the phase constant for the mode of concern is set up by equating the determinant of coefficients of Eq. (4.11) equal to zero;

$$C_{1,1}^{1,1}(\beta)C_{1,1}^{2,2}(\beta) - C_{1,1}^{1,2}(\beta)C_{1,1}^{2,1}(\beta) = 0 \qquad (4.13)$$

Compared with a characteristic equation involving a large determinant (see Eq. (3.72)), Eq. (4.13) is short and its solution can be rapidly obtained by a computer. As an example of the assessment of the computing time which could be saved using the above approximation, consider Eq. (4.13) against that obtained when using a six-term expansion for both $J_x(x)$ and $J_z(x)$. In the latter case, the number of elements involved in the resulting determinant is 144 against 4 in the former one. In view of the fact that most computer time in evaluating β is effectively associated with the computation of the elements of the determinant, by using Eq. (4.10) in contrast to employing the six-term expansion, a time saving of a factor of 36 should be achieved. This is, however, an approximate figure and there are other factors which have not been accounted for in the above assessment.

A further time saving is possible if one of the current or field distributions can be neglected. For instance, in the case of the microstrip line at low frequencies, the structure can be assumed quasi-TEM [17,18]. Thus, the transverse current is not strong and can be eliminated from calculations $(j_x(x) \approx 0)$. As a result, the characteristic equation is reduced to equation:

$$C_{1,1}^{1,2}(\beta) = 0 \qquad (4.14)$$

The above solution which makes use of one function to approximate $J_z(x)$, is often called "the zero order approximation" [19,20]. It is clear that the accuracy of β found from Eq. (4.14) depends on how closely $j_z(x)$ represents the physical distribution of the longitudinal current as well as how low the considered frequency is. In a publication by Itoh et al [19], they employ

$$j_z(x) = \frac{1}{2w}(1+|\frac{x}{w}|^3) \qquad (4.15)$$

to obtain the zero order approximation of β and use the above function together with

$$j_x(x) = \frac{1}{w}\sin\frac{\pi x}{w} \qquad (4.16)$$

to compute the propagation constant to "the first order of approximation"; see Eq. (4.13). A comparison between results generated by these two orders of approximation, Fig. 4.7, shows that the associated solutions are hardly distinguishable at low frequencies. It is worth noting here that such good accuracy (of 0 th and 1st order approximation) is problem dependent; ie: the accuracy can vary with w/h, ε_r, w/a, frequency etc. For other useful current distributions, refer to [16].

4.3 Relations among Elements of the Coefficient Matrix

As mentioned in the above section, computation of the elements of the coefficient matrix is the time consuming part in the spectral domain technique. However, a great deal of computer time can be saved if it is recognised that there is usually some kind of relations among these elements. In other words, in order to generate the whole

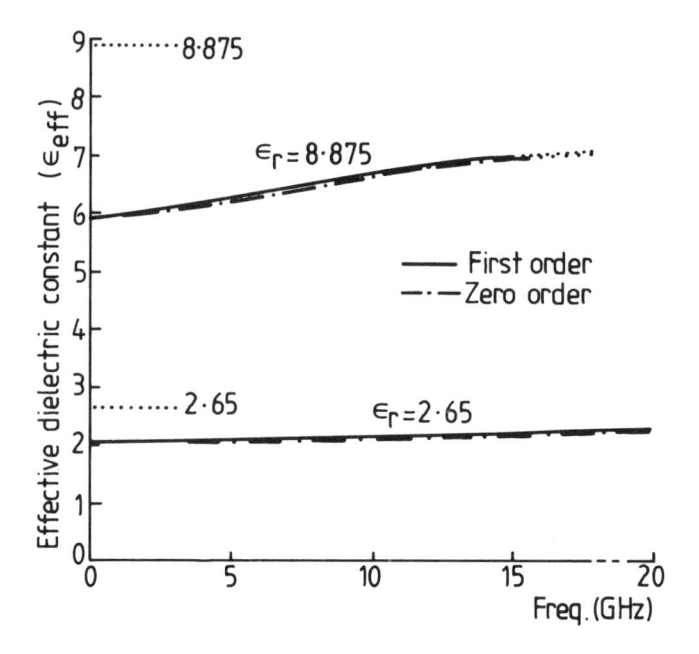

Fig. 4.7 *Effective dielectric constant of a shielded microstrip line (d=1.27 mm, 2w=1.27 mm, 2a=12.7 mm and d+h=12.7 mm) versus frequency for two different orders of the spectral domain solution and for two different dielectric loadings. (From Itoh and Mittra [19], Copyright © 1974 IEEE, reproduced with permission.)*

matrix, we actually need to compute only some of the elements. The reasons for the existence of these relations are fundamentally associated (a) with the use of Galerkin's technique and (b) with the existence of other relations among the elements of the Green's function matrix. By a special selection of the basis functions, further relations among the elements of the coefficient matrix are also feasible. Let us return to the solution of the microstrip problem now and examine the elements of its coefficient matrix.

We begin this study by writing the final equations, Eqs. (3.66) or Eq. (3.70), in a block matrix form as follows:

$$\begin{array}{c} \quad P \quad\ Q \\ Q \\ P \end{array} \left[\begin{array}{c|c} A & B \\ \hline E & F \end{array} \right] \left[\begin{array}{c} a_I \\ \vdots \\ b_I \\ \vdots \end{array} \right] = \left[\begin{array}{c} 0 \\ \vdots \\ \\ \vdots \end{array} \right] \qquad (4.17)$$

CHAP. 4 EFFICIENT COMPUTING AND APPROXIMATIONS IN ...

As seen from Eq. (4.17), the matrix of coefficients $[C(\beta)]$, Eq. (3.70), has been partitioned into four matrices A, B, E and F whose dimensions are clearly shown in Eq. (4.17).

For the even modes of the microstrip line, $J_{z,q}(x)$ is an even and $J_{x,p}(x)$ is an odd function of x. Therefore, according to Eq. (3.11.a), the Fourier transforms of $J_{z,q}(x)$ and $J_{x,p}(x)$ are purely real and purely imaginary respectively. In view of this, examination of Eqs. (3.67.b) and (3.67.c) leads to the conclusion that the two matrices B and E are symmetrical. This effect is directly associated with the application of Galerkin's technique in forming Eq. (4.17). Furthermore, inspection of the elements of matrices A and F, Eqs. (3.67.a) and Eqs. (3.67.d), reveals that

$$F = -A^T \qquad (4.18)$$

where A^T indicates the transpose of matrix A. The main reason for having the relation (4.18) lies in the fact that according to Eq. (3.37.a)

$$G_{11}(\alpha_n,\beta) = G_{22}(\alpha_n,\beta) \qquad (4.19)$$

For the odd modes where $J_{z,q}(x)$ and $J_{x,p}(x)$ are the odd and even functions of x respectively, the above conclusions still hold. There are also situations where $J_{x,p}$ and $J_{z,q}$ can be generally assumed complex. In these cases, straightforward relations as those described above may not exist, but one can still find a relation between the elements of the final matrix in order to reduce the computation effort.

As alluded earlier, a further saving of the computing time is possible if the elements of sets of basis functions approximating the field or current distributions are selected specially with the view to establishing more relations among the elements of the coefficient matrix. To give an example, we refer to the works of Jansen [1,8] in which the basis functions $J_{x,p}(x)$ and $J_{z,q}(x)$ are related by an integral expression. These basis functions and their Fourier transforms are already stated in Eqs. (4.2) and Eqs. (4.3) respectively. As can be seen from Eqs. (4.3), since

$$\tilde{J}_{x,p} = \frac{\tilde{J}_{z,p}}{\alpha_w}, \qquad (4.20)$$

provided that $P = Q$, matrices A and F will also be symmetrical. In this case, by taking into account the symmetry of matrices B and E, and the relation between A and F (Eq. (4.18)), it is possible to show that only *(3/8) (1+1/P) ×100%* of the elements of the coefficient matrix needs to be actually computed. Note that when the order of the matrix is large (P is large), the given figure is effectively 37.5%.

4.4 Alternative Spectral Domain Green's Functions

In Chapter 3, where the spectral domain formulation of the microstrip line is discussed, we arrive initially at Eqs. (3.36) which later on is presented in Eq. (3.45) as follows:

$$\begin{bmatrix} G_{11}(\alpha_n,\beta) & G_{12}(\alpha_n,\beta) \\ G_{21}(\alpha_n,\beta) & G_{22}(\alpha_n,\beta) \end{bmatrix} \begin{bmatrix} \tilde{J}_x \\ \tilde{J}_z \end{bmatrix} = \begin{bmatrix} \tilde{E}_z \\ \tilde{E}_x \end{bmatrix} \quad (4.21)$$

In the same chapter, it is mentioned that the above equation represents the coupled integral equations (3.41) in the Fourier domain. We also noted that the above equations are the basis for generating the final set of linear equations (3.66).

It will be shown later, in Chapter 6, that in the spectral domain solution of planar transmission lines with one layer of strips, Fig. 4.8, the above equation represents the typical relation between the field and the current at the interface where the conductors are laid. It is worth pointing out that for a general single substrate planar transmission line, Fig. 4.9, matrix elements $G_{11}(\alpha_n,\beta)$, $G_{12}(\alpha_n,\beta)$ etc. in Eq. (4.21) are the same as those derived previously for the microstrip line, Eqs. (3.37). As pointed out in section 3.2.4, the 4×4 matrix in Eq. (4.21) is referred to as the spectral domain Green's function matrix. Note that the dimension of this matrix is not fixed and

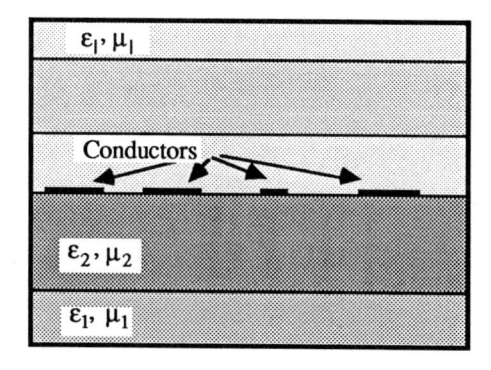

Fig. 4.8 Cross-section of a multidielectric planar transmission line with single layer of conductors.

CHAP. 4 EFFICIENT COMPUTING AND APPROXIMATIONS IN ... 67

depends on the number of conductor layers involved in a planar structure. However, since the dimension of this matrix does not alter the results of the following discussion, we concentrate on Eq. (4.21).

It is clear that Eq. (4.21) can be presented in alternative forms. For example, it can be written as:

$$\begin{bmatrix} H_{11}(\alpha_n,\beta) & H_{12}(\alpha_n,\beta) \\ H_{21}(\alpha_n,\beta) & H_{22}(\alpha_n,\beta) \end{bmatrix} \begin{bmatrix} \tilde{E}_z \\ \tilde{E}_x \end{bmatrix} = \begin{bmatrix} \tilde{J}_x \\ \tilde{J}_z \end{bmatrix} \quad (4.22)$$

or as

$$\begin{bmatrix} I_{11}(\alpha_n,\beta) & I_{12}(\alpha_n,\beta) \\ I_{21}(\alpha_n,\beta) & I_{22}(\alpha_n,\beta) \end{bmatrix} \begin{bmatrix} \tilde{J}_z \\ \tilde{E}_x \end{bmatrix} = \begin{bmatrix} \tilde{J}_x \\ \tilde{E}_z \end{bmatrix} \quad (4.23)$$

where matrix elements $H_{11}(\alpha_n,\beta)$, $H_{12}(\alpha_n,\beta)$ etc. in Eq. (4.22) and matrix elements $I_{11}(\alpha_n,\beta)$, $I_{12}(\alpha_n,\beta)$ etc. in Eq. (4.23) can be readily obtained using Eq. (4.21).

Analogous to Eq. (4.21), any one of the above two equations may be employed to set up a system of equations from which the solution of a problem of concern can be sought. Whether to employ any particular one of the Eqs. (4.21), (4.22) and (4.23) depends on the nature of the planar structure to be analysed. This is because, in some planar structures, the application of one of the mentioned equations can be associated with superior rate of convergence [13].

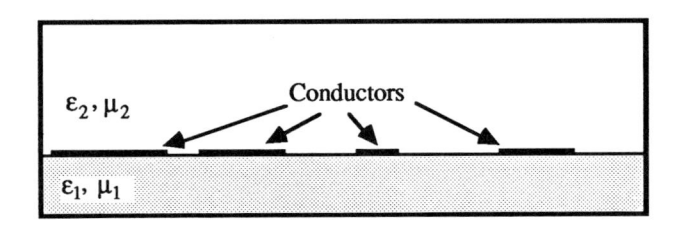

Fig. 4.9 Cross-section of a single-substrate multiconductor planar transmission line.

Experience shows that the key factor in the selection of one of the alternative forms of the spectral domain Green's functions for the solution of a problem is the metallisation distribution in that problem. In this connection, we examine the results of two solutions of a microstrip line which are obtained separately using Eq. (4.21) and Eq. (4.22).

Both solutions employ Legendre polynomials to approximate $J_z(x)$ and $E_x(x)$, and use trigonometric functions to express $J_x(x)$ and $E_z(x)$. The use of Legendre polynomials and trigonometric functions as the basis functions has already been discussed in section 4.1. As the results of this numerical experiment indicate, Fig. 4.10, solution of the microstrip line via Eq. (4.21) is considerably more efficient by comparison with that obtained by Eq. (4.22). In fact, the zero order solution *(P=0, Q=1)* achieved by Eq. (4.21) is an acceptable approximation to the actual solution. The lower numercial efficiency of the solution using Eq. (4.22) is not, however, unexpected and can be justified as follows. In the microstrip line, the electromagnetic energy is mainly confined to the region surrounding the strip. Hence at the air-dielectric interface, the strength of the electric field E_x falls sharply to a very small

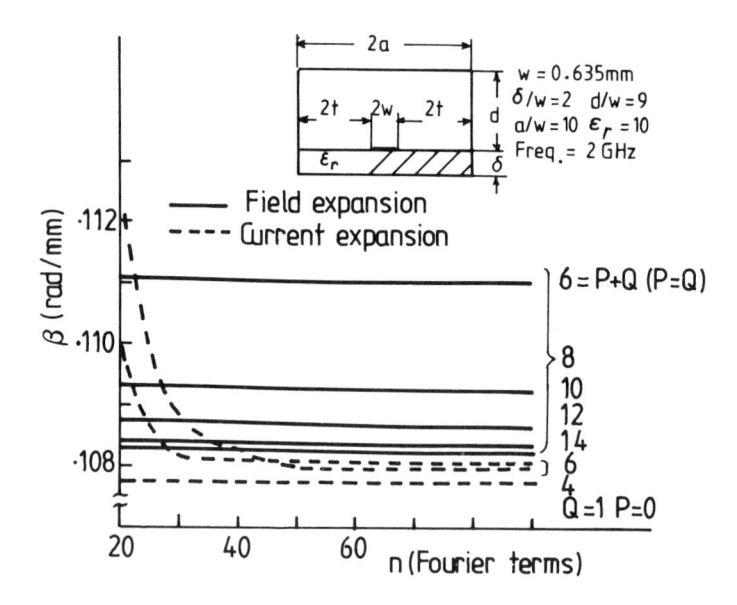

Fig. 4.10 An example of the dependence of the solution (the phase constant) on the matrix order (P+Q), the number of Fourier terms (n) and the spectral domain Green's function used in the computation.

value at some distance not very far from the edge of the strip. This vigorous behaviour of the electric field, obviously, cannot be approximated simply by a small number of basis functions. In Fig. 4.11 the computed behaviour of $E_x(x)$ for two values of $P+Q$ are shown, which substantiates the above argument.

Results of the above example and study of other investigators' works [2,13,21-23] together with some intuition lead to an empirical algorithm for the selection of an optimum spectral domain Green's function. We first examine the strip and the slot sizes in a problem of concern. If the strips are smaller than the slots, Fig. 4.12.a, it is advisable to employ Eq. (4.21). For strips larger than slot sizes, Fig. 4.12.b, Eq. (4.22) should offer a more efficient solution. Finally, if the strips and the slots are of

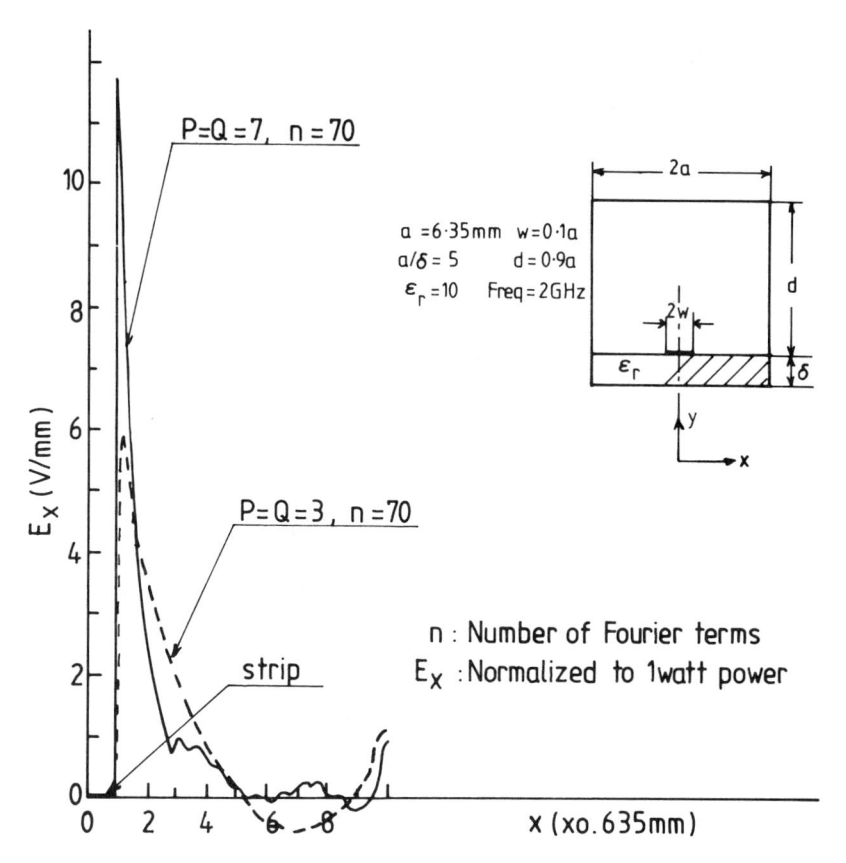

Fig. 4.11 Electric field in the x direction at the air-dielectric boundary for the fundamental mode of a microstrip line, computed for two different orders of solution using Eq. (4.22).

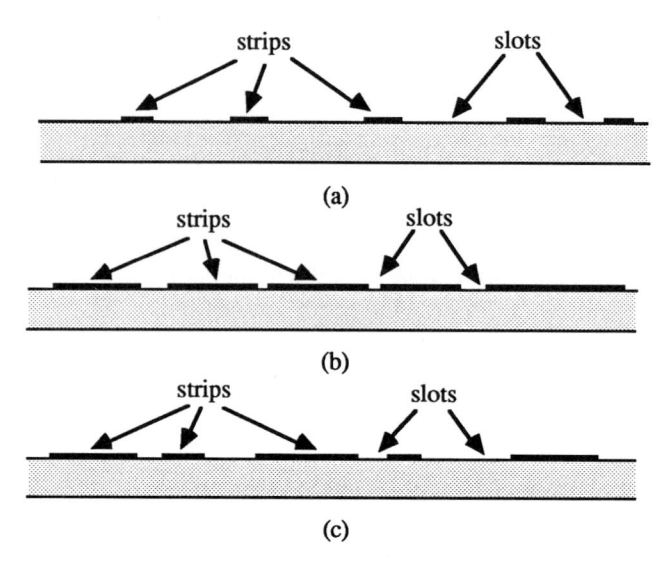

*Fig. 4.12 (a) Strips are narrower than slots,
(b) strips are wider than slots,
(c) strips and slots have comparable sizes.*

comparable sizes, Fig. 4.12.c, solution via Eq. (4.23) should require less computer time.

4.5 Acceleration of the Computation of the Matrix Elements

As mentioned in section 4.3, the time consuming part in the spectral domain solution of a planar structure is the computation of elements of the final matrix (the coefficient matrix, Eq. (3.70)). Each element is either a series or an integral (for open structures). In either case, the element needs recomputation every time that a new guess of the eigenvalue (eg: the phase constant in the case of transmission lines) is made in the iterative solution of the determinantal equation (3.72). In certain contexts, various techniques have been employed to reduce the computing time associated with calculating these elements [18,24]. We address below, one of these techniques which is particularly useful in the solution of shielded planar transmission lines. The method to be discussed is based on Hoefer's work [24], introduced originally for the analysis of E-plane circuits.

CHAP. 4 EFFICIENT COMPUTING AND APPROXIMATIONS IN ...

Let us examine equation (3.13),

$$\gamma_{i,n}^2 = \alpha_n^2 + \beta^2 - k_i^2 \qquad (4.24)$$

This expression, which invariably appears in the spectral domain Green's functions of all the planar structures, has the property that for large n, it tends to α_n,

$$\gamma_{i,n} \approx \alpha_n \qquad (4.25)$$

Now consider that each element of the final matrix is broken into two series. For instance, in the case of the microstrip line, Eq. (3.67.b) which generates some of the matrix elements can be written as follows:

$$C_{\tilde{q},q}^{1,2}(\beta) = 2\sum_{n=0}^{M} G_{12}(\alpha_n,\beta) \tilde{J}_{z,\tilde{q}}^* \tilde{J}_{z,q} + 2\sum_{M+1}^{N} G_{12}(\alpha_n,\beta) \tilde{J}_{z,\tilde{q}}^* \tilde{J}_{z,q} \qquad (4.26)$$

In view of Eq. (4.25), the series containing the higher order terms (the second series in the above equation) becomes independent of $\gamma_{i,n}$. Also, since a small variation of β has a small effect on the sum of this series as compared with the total sum, it can be assumed that this sum for values of β close to its actual value is almost constant. Thus if the iteration process starts with a close guess to β, say β', the sum of the higher order terms does not have to be computed more than once. This evidently accelerates computations. In the case of the microstrip line, the initial guess β' for the fundamental mode can be fed into computations from a quasi-TEM solution.

The important factors in the above approximation are the determination of the demarcation point of the two series M and the total number of terms N. The values of M and N depend generally on the ratio of the width of the narrowest strip or slot to the width of the box. For finlines, Fig. 4.1, Hoefer [24] proposes two empirical expressions for predicting N and M;

$$N = Integer\left[50\left(\frac{s}{a}\right)^{-0.769}\right] \qquad (4.27)$$

$$M = 3\ exp\left[-0.6\ ln\left(\frac{s}{a}\right)\right] \qquad (4.28)$$

The values of M and N found from the above equations should be sufficient in order to achieve an accuracy of better than $\pm 0.1\%$ in ε_{eff} for a guessed ε'_{eff} within $\pm 10\%$ of ε_{eff}. Parameters β and ε_{eff} are related by

$$\beta^2 = \omega^2 \mu_0 \varepsilon_0 \varepsilon_{eff} \qquad (4.29)$$

References

1. Jansen R.H., "High speed computation of single and coupled microstrip parameters including dispersion, higher order modes, loss and finite strip thickness", *IEEE Trans. Microwave Theory Tech.*, **MTT.26**, pp.75-87, 1978.

2. Jansen R.H., "Unified user-oriented computation of shielded, covered and open planar microwave and millimeter-wave transmission-line characteristics", *IEE J. Microwaves, Opt., Acoust.*, **MOA-3**, pp.14-22, 1979.

3. Lee S.W., Jones W.R. and Campbell J.J., "Convergence of numerical solution of iris type discontinuity problems", *IEEE Trans. Microwave Theory Tech.*, **MTT-19**, pp.528-536, 1971.

4. Itoh T. and Mittra R., "Relative convergence phenomenon arising in the solution of diffraction from strip grating on a dielectric slab", *Proc. Inst. Electron. Eng.*, **Vol. 59**, pp.1363-1365, 1971.

5. Hofmann H., "Relative convergence in mode matching solutions of microstrip problems", *Electron. Lett.*, **Vol. 10**, pp.126-127, 1974.

6. Pozar D.M. and Voda S.M., "A rigorous analysis of microstrip line-fed patch antenna", *IEEE Trans. Antennas and Propagation*, **AP-35**, pp.1343-1349, 1987.

7. Gradshteyn I.S. and Stegun A., *Table of Integrals, Series and Products*, Academic Press, New York, 1965.

8. Jansen R.H., "Fast accurate hybrid mode computation of nonsymmetrical coupled microstrip characteristics", 7th European Microwave Conference, pp.135-139, 1977.

9. Minor J.C. and Bolle D.M., "Modes in the shielded microstrip on a ferrite substrate transversely magnetised in the plane of the substrate", *IEEE Trans. Microwave Theory Tech.*, **MTT-19**, pp.570-577, 1971.

10. Schmidt L.P. and Itoh T., "Spectral domain analysis of dominant and higher order modes in finlines", *ibid*, **MTT-28**, pp.981-985, 1980.

11. Kantorovich L.V. and Krylov V.I., *Approximate Methods of Higher Analysis*, Interscience Publishers, New York, 1964.

12. Harrington R., "Matrix methods for field problems", *Proceedings of the IEEE*, **Vol. 55**, pp.136-149, 1967.

13. Mirshekar-Syahkal D. and Davies J.B., "Accurate solution of microstrip and coplanar structures for dispersion and for dielectric and conductor losses", *IEEE Trans. Microwave Theory Tech.*, **MTT-27**, pp.694-699, 1979.

14. Kawamoto K., Hirota K., Niizaki N., Fujiwara Y. and Ueki K., "Small size VCO module for 900 MHz band using coupled microstrip - coplanar line", *IEEE*, **MTT-S Digest**, pp.689-692, 1985.

15. Mirshekar-Syahkal D. and Davies J.B., "An accurate, unified solution to various

fin-line structures, of phase constant, characteristic impedance, and attenuation", *IEEE Trans. Microwave Theory Tech.*, **MTT-30**, pp.1854-1861, 1982.

16. Van de Capelle A.R. and Luyparet P.J., "Fundamental and higher-order modes in open microstrip lines", *Electron. Lett.*, **Vol. 9**, pp.345-346, 1973.

17. Mittra R. and Itoh T., "Analysis of microstrip transmission lines" in: *Advances in Microwaves*, **Vol. 8**, Academic Press, New York, 1974.

18. Mosig J.R. and Gardiol F.E., "Dynamical radiation model for microstrip structures" in: *Advances in Electronics and Electron Physics*, **Vol. 9**, Academic Press, New York, 1982.

19. Itoh T. and Mittra R., "A technique for computing dispersion characteristics of shielded microstrip lines", *IEEE Trans. Microwave Theory Tech.*, **MTT-22**, pp.896-897, 1974.

20. Itoh T. and Mittra R., "Spectral domain approach for calculating the dispersion characteristics of microstrip lines", *ibid*, **MTT-21**, pp.496-499, 1973.

21. Knorr J.B. and Tufekcioglu A., "Spectral-domain calculation of microstrip characteristic impedance", *ibid*, **MTT-23**, pp.725-728, 1975.

22. Janiczak B.J., "Multiconductor planar transmission line structures for high-directivity coupler applications", *IEEE*, **MTT-S Digest**, pp.215-218, 1985.

23. Knorr J.B. and Kuchler K., "Analysis of coupled slots and coplanar strips on dielectric substrate", *IEEE Trans. Microwave Theory Tech.*, **MTT-23**, pp.541-548, 1975.

24. Hoefer W.J.R., "Accelerated spectral domain analysis of E-plane circuits suitable for computer-aided design", URSI International Symposium on Electromagnetic Theory, pp.495-497, 1986.

Chapter 5

QUASI-TEM ANALYSIS BY THE SPECTRAL DOMAIN TECHNIQUE AND SOLUTION OF MULTISTRIP TRANSMISSION LINES

This chapter consists of two main parts. In the first part, a further application of the spectral domain technique is discussed. This is the quasi-TEM solution of microstrip-type structures. As in the third chapter, we introduce the solution methodology through formulating the microstrip line for which the quasi-TEM approximation under certain conditions is valid. Such treatment allows us to compare the quasi-TEM solution of the microstrip line with its full-wave solution presented in Chapter 3. One undisputed result of this comparison is that the quasi-TEM solution is more efficient numerically. But, this higher numerical efficiency is achieved at the expense of losing accuracy, and some frequency dependent information.

The rest of the chapter is devoted to a discussion on the extension of the spectral domain technique to solving multistrip single substrate transmission lines. For such structures, the Green's functions are the same as those obtained in the case of the microstrip line. It therefore remains, in the case of multistrip structures, to examine the technique of expanding components of the field and current associated with the interface where strips of conductors are laid. Although the method is specifically presented for certain planar transmission lines, it can be easily developed into a general algorithm for more complex planar structures.

5.1 Quasi-TEM Spectral Domain Formulation of the Microstrip Line

The fundamental mode of propagation on a two-conductor homogeneously-filled transmission line is TEM [1]. It is possible to show that the electromagnetic field associated with this mode is given by [1]:

$$E_t = -\nabla_t u e^{-j\beta z} \tag{5.1.a}$$

$$H_t = \left(\frac{\varepsilon}{\mu}\right)^{\frac{1}{2}} E_t \tag{5.1.b}$$

where the scalar potential function u is a function of x and y and is the solution of Laplace's equation

$$\nabla_t^2 u = 0 \tag{5.2}$$

subject to boundary conditions.

At extremely low frequencies ($f \approx 0$), the electromagnetic field in the microstrip line is essentially transverse and with a good approximation it remains so up to some higher frequencies [2,3]. Thus under TEM approximation, the fundamental mode of the microstrip line obeys the above equations.

Let us now refer to Fig. 5.1 and assume that the scalar potential, u, associated with dielectric layer i, is ϕ_i. Thus from Eq. (5.2), we have:

$$\nabla_t^2 \phi_i = 0 \qquad i=1,2 \tag{5.3}$$

Following the same procedure as that presented in section 3.2.1, the Fourier transform of the above equation is:

$$\frac{d^2}{dy^2}\tilde{\phi}_i - \alpha_n^2 \tilde{\phi}_i = 0 \tag{5.4}$$

where, according to Eq. (3.11.a), $\tilde{\phi}_i$ is given by:

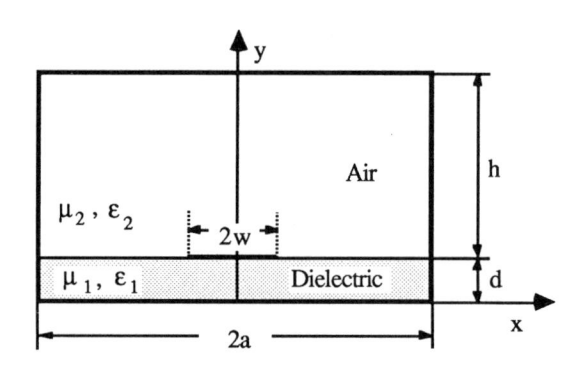

Fig. 5.1 Cross-section of a shielded microstrip line.

CHAP. 5 QUASI-TEM ANALYSIS BY ...

$$\tilde{\phi}_i = \frac{1}{2a}\int_{-a}^{+a} \phi_i e^{j\alpha_n x} dx \qquad (5.5.a)$$

The expression for the spectral parameter α_n can be obtained using the fact that ϕ_i is an even function of x and its derivative with respect to y, (which is proportional to $E_{y,i}$), must vanish at the side walls of the microstrip shield. Using these conditions, we obtain:

$$\alpha_n = (n+\tfrac{1}{2})\frac{\pi}{a} \qquad n=0,1,2,... \qquad (5.5.b)$$

Now assuming that the shield is at zero potential, it is easy to show that the spectral domain potential functions satisfying this condition as well as Eq. (5.4) are:

$$\tilde{\phi}_1(\alpha_n,y) = A_{1,n}\sinh(\alpha_n y) \qquad 0<y<d \qquad (5.6.a)$$

$$\tilde{\phi}_2(\alpha_n,y) = B_{2,n}\sinh[\alpha_n(h+d-y)] \qquad d<y<d+h \qquad (5.6.b)$$

The given potentials are fully determined when $A_{1,n}$ and $B_{2,n}$ are known. To this end, we make use of the boundary conditions at the air-dielectric boundary $(y=d)$. These conditions in the space domain are:

$$E_{x,2} - E_{x,1} = 0 \qquad (5.7.a)$$

$$\varepsilon_2 E_{y,2} - \varepsilon_1 E_{y,1} = \begin{cases} \sigma(x)e^{-j\beta z} & |x|<w \\ 0 & w<|x|<a \end{cases} \qquad (5.7.b)$$

$$\phi_1 = \phi_2 = \begin{cases} V_0 & |x|<w \\ V(x) & w<|x|<a \end{cases} \qquad (5.7.c)$$

and in the Fourier domain in terms of the potential functions are:

$$\tilde{\phi}_2 - \tilde{\phi}_1 = 0 \qquad (5.8.a)$$

$$\varepsilon_2 \frac{d\tilde{\phi}_2}{dy} - \varepsilon_1 \frac{d\tilde{\phi}_1}{dy} = \tilde{\sigma} \qquad (5.8.b)$$

$$\tilde{\phi}_1 = \tilde{\phi}_2 = \tilde{V}_0 + \tilde{V} \qquad (5.8.c)$$

In the above equations $\sigma(x)$ is the surface charge density on the strip and $V(x)$ is the potential distribution at the interface excluding the strip. The strip itself is assumed to be at potential V_0. It should be mentioned that the Fourier transforms of V_0 and $V(x)$ are given by:

$$\tilde{V}_0 = \frac{1}{2a} \int_{-w}^{+w} V_0 e^{j\alpha_n x} dx \qquad (5.9.a)$$

$$\tilde{V} = \frac{1}{2a} \left[\int_{-a}^{-w} V(x) e^{j\alpha_n x} dx + \int_{w}^{a} V(x) e^{j\alpha_n x} dx \right] \qquad (5.9.b)$$

Substituting Eqs. (5.6) in the boundary conditions (5.8.a) and (5.8.b) leads to the following relations between $A_{1,n}$ and $B_{1,n}$:

$$A_{1,n} \sinh(\alpha_n d) - B_{2,n} \sinh(\alpha_n h) = 0 \qquad (5.10.a)$$

$$\varepsilon_1 \alpha_n A_{1,n} \cosh(\alpha_n d) + \varepsilon_2 \alpha_n B_{2,n} \cosh(\alpha_n h) = \tilde{\sigma} \qquad (5.10.b)$$

From these equations $A_{1,n}$ and $B_{2,n}$ in terms of the Fourier transform of σ can be obtained. For instance,

$$B_{2,n} = \frac{\tilde{\sigma}}{\alpha_n \sinh(\alpha_n h)[\varepsilon_1 \coth(\alpha_n d) + \varepsilon_2 \coth(\alpha_n h)]} \qquad (5.11)$$

But, since the charge distribution on the strip is not yet known, $A_{1,n}$ and $B_{2,n}$ are not fully determined. With the help of the final boundary condition, Eq. (5.8.c), we will be able to find the charge distribution and hence to complete the solution.

Substituting $B_{2,n}$ in Eq. (5.6.b) and inserting the resulting expression in Eq. (5.8.c), we obtain:

$$G(\alpha_n)\tilde{\sigma} = \tilde{V}_0 + \tilde{V} \qquad (5.12)$$

where

$$G(\alpha_n) = [\alpha_n(\varepsilon_1 \coth(\alpha_n d) + \varepsilon_2 \coth(\alpha_n h))]^{-1} \qquad (5.13)$$

Equation (5.12) is similar to Eqs. (3.36) derived in the full-wave spectral domain analysis of the microstrip line. Thus, it represents the Fourier transform of a Fredholm's integral equation in the space domain. However, in contrast to Eqs.

(3.36), Eq. (5.12) is much simpler and contains no implicit unknown variables such as β that exist in Eqs. (3.36). Also it can be observed from Eq. (5.13) that under quasi-TEM approximation, the spectral domain Green's function of the problem is not dyadic.

Similar to Eqs. (3.36) in the full-wave case, Eq. (5.12) is the final step in the quasi-static spectral domain formulation of the microstrip line. To solve this equation, the Galerkin technique may be employed. This technique has already been introduced in section 3.3.2. To apply it, we select a set of basis functions $\{\sigma_p(x), p=1,2,...\}$ and write $\sigma(x)$ as follows:

$$\sigma(x) = \sum_{p=1}^{P} a_p \sigma_p(x) \tag{5.14}$$

We remember that the basis functions must be chosen so that they satisfy condition (5.7.b), ie:

$$\sigma_p(x) = 0 \qquad\qquad w < |x| < a \tag{5.15}$$

and other relevant criteria discussed in section 4.1. Substituting the Fourier transform of Eq. (5.14) in Eq. (5.12) and taking the inner product of both sides of the resulting equation with the basis functions according to Eq. (3.74) lead to the following system of equations:

$$\sum_{p=1}^{P} C_{p',p} a_p = H_{p'} \qquad\qquad p' = 1,2,... \tag{5.16}$$

where

$$C_{p',p} = \sum_{n=-\infty}^{+\infty} G(\alpha_n) \tilde{\sigma}_{p'}^* \tilde{\sigma}_p \tag{5.17.a}$$

$$H_{p'} = \sum_{n=-\infty}^{+\infty} \tilde{V}_0 \tilde{\sigma}_{p'}^* \tag{5.17.b}$$

In forming Eq. (5.16), the inner products

$$<\tilde{V}, \tilde{\sigma}_{p'}> \qquad\qquad p' = 1,2,... \tag{5.18}$$

do not contribute to this equation. This can be proved using Parseval's identity, Eq. (3.68), which turns the above expression into an integral in the space domain as follows:

$$<\tilde{V}, \tilde{\sigma}_p> = \frac{1}{2a}\int_{-a}^{+a} V(x)\sigma_p^*(x)dx \qquad (5.19)$$

The value of this integral is zero, because

$$V(x) = 0 \qquad |x|<w \qquad (5.20.a)$$

$$\sigma(x) = 0 \qquad w<|x|<a \qquad (5.20.b)$$

It is very clear that Eq. (5.16) has a straightforward solution yielding all the expansion parameters a_p. The solution is achieved by inverting a matrix and in contrast to solution of Eqs. (3.66) involves no root finding process. Thus, it is numerically more efficient and requires less computer time. Furthermore, the quasi-TEM assumption has reduced significantly the mathematical effort in formulating the problem. However, as alluded earlier, these computational and analytical advantages are gained at the expense of losing accuracy in determining frequency dependent information.

For a further example giving more insight into the above technique, the reader is referred to a work by Itoh et al [4]. In this work, the coupled suspended microstrip lines with tuning septums, Fig. 5.2, has been analysed in detail.

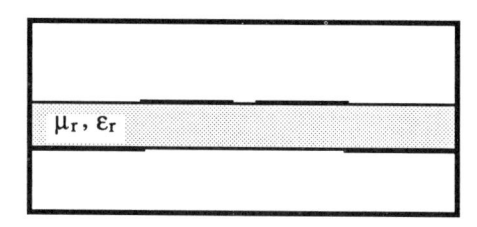

Fig. 5.2 Cross-section of a coupled microstrip line with septums.

5.2 Spectral Domain Solution of Multistrip Single-Substrate Transmission Lines

The generic cross-sections of multistrip single-substrate transmission lines are shown in Fig. 5.3. Many planar coupler structures fall within one of these categories (for example see [5,6]). Thus their analyses basically obey the same rules.

CHAP. 5 QUASI-TEM ANALYSIS BY ...

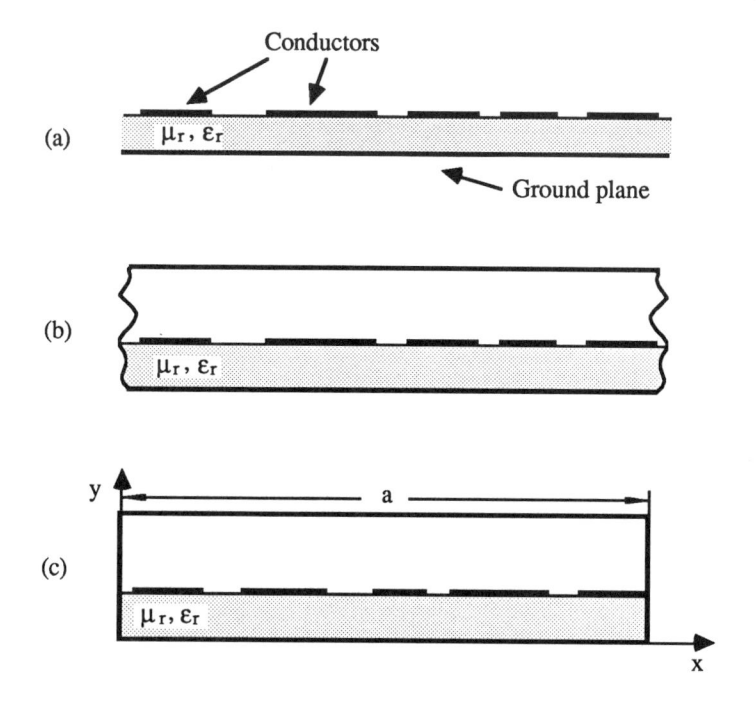

(a) [Conductors, μ_r, ε_r, Ground plane]

(b) [μ_r, ε_r]

(c) [y, a, μ_r, ε_r, x]

Fig. 5.3 Cross-sections of multistrip single substrate planar transmission lines; (a) open, (b) laterally open and (c) shielded structure.

As far as the spectral domain solution of a multistrip single-substrate planar transmission line is concerned, it is very similar to that of the microstrip line. In fact, the associated spectral domain Green's functions in both cases are the same. Thus, Eqs. (3.36) in the full-wave and Eq. (5.12) in the quasi-TEM analysis are directly applicable to the solution of shielded multistrip single-substrate planar lines. What eventually makes the solution of a multistrip structure different from that of a structure with one strip is the way that the current or field components are defined in order to solve the integral equations. The following examples help to clarify this matter.

5.2.1 Asymmetric Coupled Microstrip Lines

Consider the asymmetric coupled microstrip line shown in Fig. 5.4 and assume that the full-wave analysis of this structure is required. This can be achieved by using Eqs. (3.36) and converting them into a set of homogeneous linear equations. For this

CHAP. 5 QUASI-TEM ANALYSIS BY ...

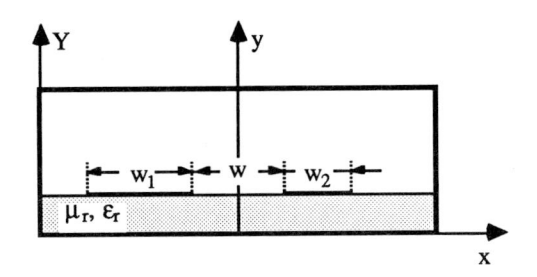

Fig. 5.4 Cross-section of an asymmetrical shielded microstrip line.

purpose, the method proposed in section 3.3.2.2 for the microstrip line is followed below.

Since the two strips involved in the present problem are of unequal sizes, we initially write $J_x(x)$ and $J_z(x)$ as follows:

$$J_x(x) = J_{x,1}(x) + J_{x,2}(x) \tag{5.21.a}$$

$$J_z(x) = J_{z,1}(x) + J_{z,2}(x) \tag{5.21.b}$$

where $J_{x,1}(x)$, $J_{z,1}(x)$ and $J_{x,2}(x)$, $J_{z,2}(x)$ are the components of currents on the first and the second strip respectively. The above equations are now in appropriate forms and each of the unknown elements in these equations may be expressed in terms of a set of suitable basis functions. Hence, in expanded forms, these equations are given as follows:

$$J_x(x) = \sum_{q=1}^{Q} b_q g_q(x) + \sum_{s=1}^{S} d_s i_s(x) \tag{5.22.a}$$

$$J_z(x) = \sum_{p=1}^{P} a_p f_p(x) + \sum_{r=1}^{R} c_r h_r(x) \tag{5.22.b}$$

where the basis functions $f_p(x)$ and $g_q(x)$ both equal zero for x outside strip 1 and the basis functions $h_r(x)$ and $i_s(x)$ both equal zero for x outside strip 2.

Typical expressions for these expansion functions can be found in the works of

CHAP. 5 QUASI-TEM ANALYSIS BY ...

Jansen [7] and Kitlinski et al [8]. Using the methodology of Kitlinski et al, the above basis functions can be written as follows:

$$f_p(x) = [1 - \theta_1^2(x)]^{-\frac{1}{2}} \cos[p\pi\eta_1(x)] \qquad (5.23.a)$$

$$-\frac{w}{2} > x > -\frac{w}{2} - w_1$$

$$g_q(x) = \sin[q\pi\eta_1(x)] \qquad (5.23.b)$$

$$h_r(x) = [1-\theta_2^2(x)]^{-\frac{1}{2}} \cos[r\pi\eta_2(x)] \qquad (5.23.c)$$

$$\frac{w}{2} < x < \frac{w}{2} + w_2$$

$$i_s(x) = \sin[s\pi\eta_2(x)] \qquad (5.23.d)$$

where

$$\theta_l(x) = \frac{2x + (-1)^{l+1}(w + w_l)}{w_l} \qquad (5.23.e)$$

$$l = 1, 2$$

$$\eta_l(x) = \frac{2x + (-1)^{l+1}(w + 2w_l)}{2w_l} \qquad (5.23.f)$$

With reference to section 4.1, it is clear that the above functions have closed-form Fourier transforms and they also meet the criteria set out for numerically efficient basis functions.

Having defined the basis functions, the next step in the solution is to insert the Fourier transforms of the expanded forms of currents, Eqs. (5.22), into Eqs. (3.36). The result is as follows:

$$\tilde{E}_z = \sum_q b_q \tilde{g}_q G_{11}(\alpha_n, \beta) + \sum_s d_s \tilde{i}_s G_{11}(\alpha_n, \beta) + \sum_p a_p \tilde{f}_p G_{12}(\alpha_n, \beta) \qquad (5.24.a)$$

$$+ \sum_r c_r \tilde{h}_r G_{12}(\alpha_n, \beta)$$

$$\tilde{E}_x = \sum_q b_q \tilde{g}_q G_{21}(\alpha_n, \beta) + \sum_s d_s \tilde{i}_s G_{21}(\alpha_n, \beta) + \sum_p a_p \tilde{f}_p G_{22}(\alpha_n, \beta) \qquad (5.24.b)$$

$$+ \sum_r c_r \tilde{h}_r G_{22}(\alpha_n, \beta)$$

Using the Galerkin technique, the above equations can be turned into a homogeneous system of equations. The procedure is already discussed for the case of the microstrip line and hence is not repeated again.

Comparing Eqs. (5.24) with Eqs. (3.73), it is clear that the computing time increases with the number of strips and can therefore be prohibitive for structures with several strips. In such cases, it is appropriate to use approximated forms for currents or fields in order to reduce the computing time to a reasonable level. This approximation is explained in detail in section 4.2. In this connection, the work reported in [9] on the analysis of shielded coupled coplanar waveguides is also worth consulting.

One last important point in respect of the solution of the shielded asymmetric coupled microstrip line is the way that parameter α_n is defined. Unlike the microstrip line, asymmetric structures do not enjoy any electric or magnetic wall symmetry. Thus α_n cannot be expressed by Eq. (3.9) or Eq. (3.5). For such structures, the expression for α_n is given by:

$$\alpha_n = \frac{n\pi}{a} \qquad n=0,1,2,... \qquad (5.25)$$

where a is the whole width of the shield. In these cases, the y-axis must be placed along one of the side walls, Fig. 5.3.c. This ensures that the boundary conditions for the electric and magnetic field components are automatically met when they are written in the Fourier series.

5.2.2 Multiconductor Microstrip Lines

A very interesting application of the spectral domain technique in the analysis of multistrip structures can be found in a paper by Farr et al [10]. This work addresses the problem of coupling of energy among interconnecting striplines in high-speed VLSI circuits. The proposed solution for this problem makes use of the coupled mode theory while the necessary characteristics of various propagating modes are computed using the spectral domain technique. The structure treated in the paper consists of five microstrip lines etched on a single substrate, Fig. 5.5. Since the structure enjoys a symmetry, one can take advantage of this symmetry and as in the case of microstrip line, the modes can be specified as even or odd. Irrespective of the mode of concern, however, the expansion technique used in writing Eqs. (5.22) can be generalised in order to expand $J_x(x)$ and $J_z(x)$ for the present problem:

CHAP. 5 QUASI-TEM ANALYSIS BY ...

$$J_x(x) = \sum_{l=1}^{5} \sum_{p=1}^{P} a_{l,p} f_{l,p}(x) \qquad (5.26.\text{a})$$

$$J_z(x) = \sum_{l=1}^{5} \sum_{q=1}^{Q} b_{l,q} g_{l,q}(x) \qquad (5.26.\text{b})$$

where $f_{l,p}(x)$ and $g_{l,q}(x)$ are the basis functions. In [10], these basis functions are formed using the following functions:

$$\xi_{e,p}(x) = S^{-1}(x)\cos[(p-1)\pi(x/w-1)] \qquad (5.27.\text{a})$$

$$\eta_{e,p}(x) = S(x)\sin[(p-\tfrac{1}{2})\pi(x/w+1)] \qquad (5.27.\text{b})$$

$$\xi_{o,p}(x) = S^{-1}(x)\cos[(p-\tfrac{1}{2})\pi(x/w+1)] \qquad (5.27.\text{c})$$

$$\eta_{o,p}(x) = S(x)\sin[p\pi(x/w-1)] \qquad (5.27.\text{d})$$

where

$$S(x) = [1-(\tfrac{x}{w})^2]^{\tfrac{1}{2}} \qquad (5.27.\text{e})$$

Note that the above functions are assumed zero for $|x| > w$. For instance, for evaluating the characteristics of the even modes, the appropriate basis functions built up from the above expressions are as follows:

$$f_{1,p}(x) = \eta_{o,p}(x) \qquad (5.28.\text{a})$$

$$f_{2,p}(x) = \eta_{o,p}(x+s) + \eta_{o,p}(x-s) \qquad (5.28.\text{b})$$

$$f_{3,p}(x) = \eta_{e,p}(x+s) - \eta_{e,p}(x-s) \qquad (5.28.\text{c})$$

$$f_{4,p}(x) = \eta_{o,p}(x+2s) + \eta_{o,p}(x-2s) \qquad (5.28.\text{d})$$

$$f_{5,p}(x) = \eta_{e,p}(x+2s) - \eta_{e,p}(x-2s) \qquad (5.28.\text{e})$$

$$g_{1,q}(x) = \xi_{e,q}(x) \qquad (5.29.\text{a})$$

$$g_{2,q}(x) = \xi_{e,q}(x+s) + \xi_{e,q}(x-s) \tag{5.29.b}$$

$$g_{3,q}(x) = \xi_{o,q}(x+s) - \xi_{o,q}(x-s) \tag{5.29.c}$$

$$g_{4,q}(x) = \xi_{e,q}(x+2s) + \xi_{e,q}(x-2s) \tag{5.29.d}$$

$$g_{5,q}(x) = \xi_{o,q}(x+2s) - \xi_{o,q}(x-2s) \tag{5.29.e}$$

where

$$s = 2(w+w') \tag{5.30}$$

Note that in the above expressions, $\eta_{o,p}(x)$ is zero outside the first strip, $\eta_{o,p}(x-s)$ is zero outside the second strip, $\eta_{o,p}(x+s)$ is zero outside the third strip and so on. Also note that in the case of using the above expansion functions, the index 5 in Eqs. (5.26) does not represent the number of strips involved in the problem.

From this stage onwards, the solution of the problem is very similar to that explained in the previous section. It is clear that the above method of generating the basis functions using the kernel functions (5.27) can be readily extended to planar structures with many strips of equal sizes. Also by comparing Eqs. (5.27) with (5.23), it is easily recognisable that by some modification to Eqs. (5.27), they can be converted to Eqs. (5.23).

From the study of the two examples presented in this and previous sections, the main conclusion is that multistrip planar structures are generally solvable using the spectral domain technique. However, particular attention must be paid to the choice of the basis functions as the computing time can be prohibitively large for unwisely chosen basis functions.

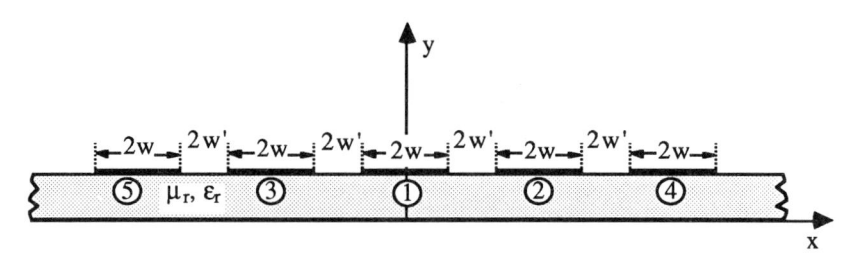

Fig. 5.5 Cross-section of five coupled microstrip lines. (From Farr et al [10], Copyright © 1986 IEEE, reproduced with permission.)

References

1. Collin R.E., *Field Theory of Guided Waves*, McGraw Hill, 1966.

2. Mittra R. and Itoh T., "Analysis of microstrip transmission lines" in: *Advances in Microwaves,* **Vol. 8**, Academic Press, New York, 1974.

3. Mosig J.R. and Gardiol F.E., "Dynamical radiation model for microstrip structures", in: *Advances in Electronics and Electron Physics,* **Vol. 9**, Academic Press, New York, 1982.

4. Itoh T. and Herbert A.S., "A generalised spectral domain analysis for coupled suspended microstriplines with tuning septums", *IEEE Trans. Microwave Theory Tech.,* **MTT-26**, pp.820-826, 1978.

5. Janiczak B.J., "Multiconductor planar transmission-line structures for high-directivity coupler applications", *IEEE,* **MTT-S Digest**, pp.215-218, 1985.

6. Tajima Y. and Kamihashi S., "Multiconductor couplers", *IEEE Trans. Microwave Theory Tech.,* **MTT-26**, pp.795-801, 1978.

7. Jansen R.H., "Fast accurate hybrid mode computation of nonsymmetrical coupled microstrip characteristics", *Proc. 7th European Microwave Conference,* pp.135-139, 1977.

8. Kitlinski M. and Janiczak B., "Dispersion characteristics of asymmetric coupled slot lines on dielectric substrate", *Electron. Lett.,* **Vol. 19**, pp.91-92, 1983.

9. Mirshekar-Syahkal D., "Dispersion in shielded coupled coplanar waveguides", *Electron. Lett.,* **Vol. 22**, pp.358-360, 1986.

10. Farr E.G., Chan C.H. and Mittra R., "A frequency-dependent coupled-mode analysis of multiconductor microstrip lines with application to VLSI interconnection problems", *IEEE Trans. Microwave Theory Tech.,* **MTT-34**, pp.307-310, 1986.

Chapter 6

SPECTRAL DOMAIN SOLUTION OF MULTILAYER MULTICONDUCTOR PLANAR TRANSMISSION LINES

An increasing use of multilayer multiconductor planar structures in the development of many microwave devices is clearly evident in recent publications. Layers of suitable materials may be added to improve the performance of a device or they may be required as essential building blocks in the design of a component. For example, in microstrip couplers, Fig. 6.1.a, extra layers of dielectrics may be used to enhance the directivity [1-4], whereas the structure of a finline isolator [5,6] consists in principle of four layers, of which one is ferrite, Fig. 6.1.b. As can be seen in Fig. 6.1, the coupler and the isolator in the above examples are multilayer multiconductor planar structures. Multilayer multiconductor structures are also reported to be used in high-speed digital integrated circuits [7,8].

A multilayer multiconductor structure is generally referred to a structure whose conductors are distributed at various layers. An example is shown in Fig. 6.2. But more practical types of multilayer multiconductor structures are those whose conductors are coplanar. In Fig. 6.3, an instance of one such structure is depicted. Multilayer structures with coplanar conductors are easier to make as well as being more convenient to analyse.

The main objective of this chapter is to show how the spectral domain technique can be applied in order to solve multilayer multiconductor planar structures. For this purpose, we begin the discussion by considering the solution of transmission lines consisting of many dielectric layers, but having conductors distributed at one dielectric interface only, Fig. 6.3 and Fig. 6.4. This solution is then generalised to the case of more complex transmission lines in which conductors are arbitrarily distributed at various interfaces, Fig. 6.2.

As will be demonstrated shortly, one way of solving multilayer structures by the

CHAP. 6 SPECTRAL DOMAIN SOLUTION OF MULTILAYER . . . 90

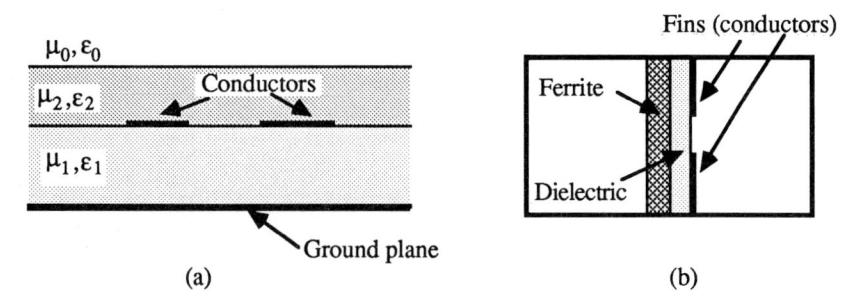

Fig. 6.1 (a) Cross-section of a microstrip coupler with dielectric overlay and (b) cross-section of a finline isolator.

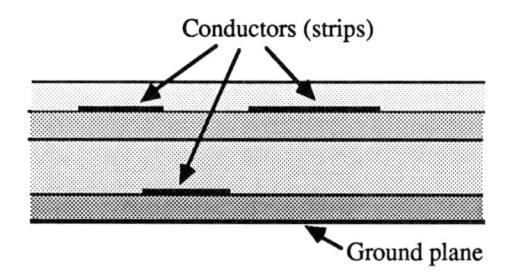

Fig. 6.2 Cross-section of a multilayer multiconductor planar transmission line.

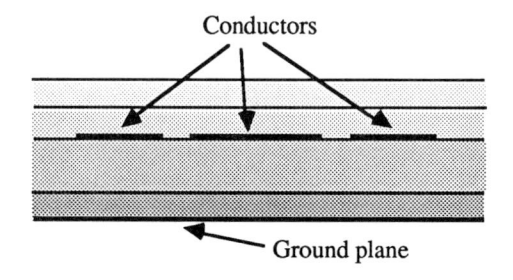

Fig. 6.3 Cross-section of a multilayer tranmission line with coplanar conductors.

CHAP. 6 SPECTRAL DOMAIN SOLUTION OF MULTILAYER ...   91

spectral domain technique is to use a "transfer matrix". Introduction of this matrix allows us to relate the coefficients of the potential functions of one layer to those of subsequent ones. The transfer matrix approach, originally given in [9], offers a simple practical tool for the generation of the Green's functions of multilayer structures in the Fourier domain. The notion of the transfer matrix is applicable under both full-wave and quasi-TEM assumptions.

In this chapter, when dealing with a general multilayer multiconductor structure, it will be shown that the order of the Green's function matrix $[G(\alpha_n,\beta)]$ depends on the number of layers of conductors only. In fact, each conductor layer increases the matrix order by two.

Also in this chapter, we briefly mention other techniques of formulating the Green's function of a multilayer multiconductor structure. One of these techniques, the spectral domain immittance approach, reported by Itoh [10] will be examined in some detail.

Explicit expressions for the Green's functions of certain multilayer multiconductor structures including finlines, coupled microstrip finlines and coplanar waveguides are given in this chapter. These expressions can be immediately used for the analysis and design of those as well as many other structures.

6.1 Spectral Domain Solution of Multilayer Transmission Lines with Coplanar Conductors Using Transfer Matrix Approach

The generic cross-section of a transmission line with multilayer dielectric and coplanar conductors is shown in Fig. 6.4. As illustrated in this figure, the structure is assumed to be shielded. However, as explained in the case of microstrip line, the spectral domain formulation of a shielded structure can be easily modified for the solution of the same structure when it is considered open or laterally open.

As mentioned in section 2.5, structures of the type shown in Fig. 6.4 generally support hybrid modes. Thus the field in each dielectric region can be regarded as the superposition of TE and TM modes. In other words, the field of each region can be expressed in terms of two scalar potential functions. In connection with the full-wave solution of the microstrip line, this is already explained fully in Chapter 3.

As in the case of the microstrip line, the principal objective in the process of the spectral domain solution of the structure shown in Fig. 6.4 is to establish a relationship among the Fourier transforms of current and field distributions $J_x(x)$,

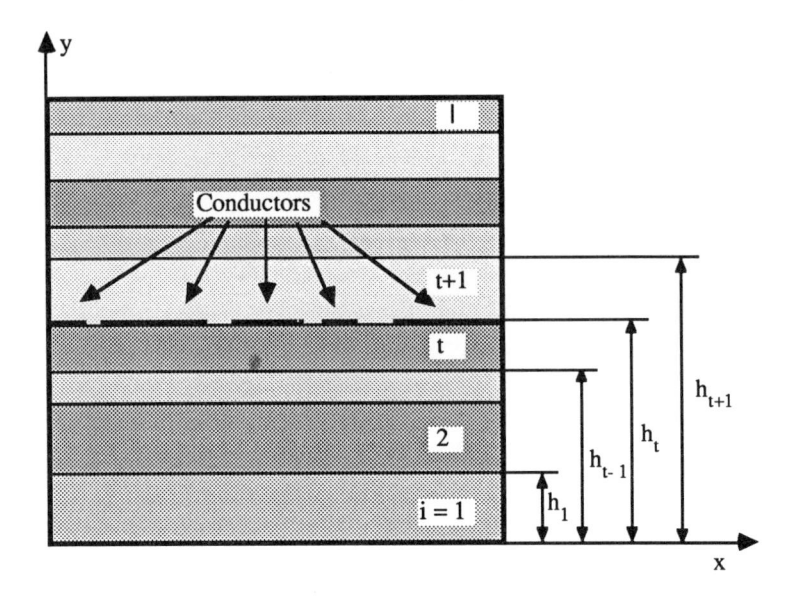

Fig. 6.4 Cross-section of a shielded multilayer transmission line with coplanar conductors.

$J_z(x)$, $E_x(x)$ and $E_z(x)$ associated with the interface where strips are laid. In mathematical terms, we are primarily concerned with seeking the Green's function, $[G(\alpha_n,\beta)]$, of the problem. As will be shown below, the Green's function of a multilayer planar structure can be systematically obtained using a transfer matrix. In the derivation of this matrix, we assume that the dielectrics involved in the structure are lossless, homogeneous and isotropic. Furthermore, it is assumed that the strips and the structure shield are made of a perfect conductor.

We begin the derivation of the transfer matrix by assuming that the longitudinal field components of each dielectric region can be written as follows (see section 2.5):

$$E_{z,i} = j\frac{k_i^2-\beta^2}{\beta}\phi_i e^{-j\beta z} \tag{6.1.a}$$

$$H_{z,i} = j\frac{k_i^2-\beta^2}{\beta}\psi_i e^{-j\beta z} \tag{6.1.b}$$

where i is the dielectric index. In the above equations, all the functions and parameters are similar to those defined in Chapter 3.

CHAP. 6 SPECTRAL DOMAIN SOLUTION OF MULTILAYER...

Now considering the Fourier transform defined by Eq. (3.11.a) and pursuing the same procedure introduced in section 3.2.1, we can readily write the spectral domain representation of ϕ_i and ψ_i. These are given as follows:

$$\tilde{\phi}_i(\alpha_n, y) = A_{i,n} \sinh(\gamma_{i,n} y) + B_{i,n} \cosh(\gamma_{i,n} y) \qquad (6.2.a)$$
$$i \leq t$$
$$\tilde{\psi}_i(\alpha_n, y) = C_{i,n} \sinh(\gamma_{i,n} y) + D_{i,n} \cosh(\gamma_{i,n} y) \qquad (6.2.b)$$

$$\tilde{\phi}_i(\alpha_n, y) = A_{i,n} \sinh[\gamma_{i,n}(h_i - y)] + B_{i,n} \cosh[\gamma_{i,n}(h_i - y)] \qquad (6.2.c)$$
$$i \geq t+1$$
$$\tilde{\psi}_i(\alpha_n, y) = C_{i,n} \sinh[\gamma_{i,n}(h_i - y)] + D_{i,n} \cosh[\gamma_{i,n}(h_i - y)] \qquad (6.2.d)$$

As can be inferred from the above expressions, potential functions associated with dielectrics above the plane of strips $(i \geq t+1)$ are represented differently compared with those associated with dielectrics below the plane of strips $(i \leq t)$. This is done only to simplify the mathematical formulation. In the above equations

$$\gamma_{i,n}^2 = \alpha_n^2 + \beta^2 - k_i^2 \qquad (6.3)$$

and $A_{i,n}$, $B_{i,n}$, $C_{i,n}$ and $D_{i,n}$ are coefficients of the potential functions. These coefficients as well as the phase constant β, are the main unknowns of the problem.

In order to set up a transfer matrix equation relating the potential function coefficients of one layer to those of the subsequent one, the boundary conditions between the two layers are first evoked. With reference to Eqs. (3.23), these conditions in the Fourier domain for two consecutive layers of i and $i+1$ are:

$$\tilde{E}_{z,i} = \tilde{E}_{z,i+1} \qquad (6.4.a)$$
$$\tilde{E}_{x,i} = \tilde{E}_{x,i+1} \qquad (6.4.b)$$
$$\tilde{H}_{z,i} = \tilde{H}_{z,i+1} \qquad (6.4.c)$$
$$\tilde{H}_{x,i} = \tilde{H}_{x,i+1} \qquad (6.4.d)$$

As mentioned in sections 2.3 and 3.2.2, since the field in each dielectric region is the solution of Maxwell's equations, satisfaction of the above boundary conditions is sufficient to describe the field completely. For this reason, the other two boundary conditions are excluded from the solution.

CHAP. 6 SPECTRAL DOMAIN SOLUTION OF MULTILAYER ...

Substituting the Fourier transforms of the potential functions, Eqs. (6.2), in the Fourier domain expressions of the field components, Eqs. (3.19), and inserting the resulting equations in Eqs. (6.4) lead to the desired equation:

$$\begin{bmatrix} A_{i+1,n} \\ B_{i+1,n} \\ C_{i+1,n} \\ D_{i+1,n} \end{bmatrix} = [T_{i+1,i}] \begin{bmatrix} A_{i,n} \\ B_{i,n} \\ C_{i,n} \\ D_{i,n} \end{bmatrix} \qquad (6.5)$$

where

$$[T_{i+1,i}] = [\gamma_{i+1} u_i]^{-1} [\gamma_i v_i] \qquad (6.6)$$

and

$$[\gamma_i v_i] = \begin{bmatrix} (k_i^2-\beta^2)s_v & (k_i^2-\beta^2)c_v & 0 & 0 \\ 0 & 0 & (k_i^2-\beta^2)s_v & (k_i^2-\beta^2)c_v \\ -j\alpha_n s_v & -j\alpha_n c_v & \mp\dfrac{\omega\mu_i\gamma_{i,n}}{\beta}c_v & \mp\dfrac{\omega\mu_i\gamma_{i,n}}{\beta}s_v \\ \pm\dfrac{\omega\varepsilon_i\gamma_{i,n}}{\beta}c_v & \pm\dfrac{\omega\varepsilon_i\gamma_{i,n}}{\beta}s_v & -j\alpha_n s_v & -j\alpha_n c_v \end{bmatrix} \qquad (6.7)$$

$$[\gamma_{i+1} u_i] = \begin{bmatrix} (k_{i+1}^2-\beta^2)s_u & (k_{i+1}^2-\beta^2)c_u & 0 & 0 \\ 0 & 0 & (k_{i+1}^2-\beta^2)s_u & (k_{i+1}^2-\beta^2)c_u \\ -j\alpha_n s_u & -j\alpha_n c_u & \mp\dfrac{\omega\mu_{i+1}\gamma_{i+1,n}}{\beta}c_u & \mp\dfrac{\omega\mu_{i+1}\gamma_{i+1,n}}{\beta}s_u \\ \pm\dfrac{\omega\varepsilon_{i+1}\gamma_{i+1,n}}{\beta}c_u & \pm\dfrac{\omega\varepsilon_{i+1}\gamma_{i+1,n}}{\beta}s_u & -j\alpha_n s_u & -j\alpha_n c_u \end{bmatrix}$$

$$(6.8)$$

CHAP. 6 SPECTRAL DOMAIN SOLUTION OF MULTILAYER...

In the above matrices

$$s_v = \sinh(\gamma_{i,n} v_i) \qquad s_u = \sinh(\gamma_{i+1,n} u_i)$$
$$c_v = \cosh(\gamma_{i,n} v_i) \qquad c_u = \cosh(\gamma_{i+1,n} u_i)$$

where v_i and u_i are given by

$$\begin{cases} v_i = h_i \\ u_i = h_i \end{cases} \qquad \text{for } i \leq t \qquad (6.9.\text{a})$$

$$\begin{cases} v_i = 0 \\ u_i = h_{i+1} - h_i \end{cases} \qquad \text{for } i \geq t+1 \qquad (6.9.\text{b})$$

Also in the same matrices, the upper signs are for $i \geq t+1$ and the lower signs are for $i \leq t$. Note that in the derivation of Eq. (6.5), the Fourier transforms of the field components, Eqs. (3.19), were initially multiplied by j/β in order to make them compatible with the Fourier transforms of Eqs. (6.1).

The 4×4 matrix $[T_{i+1,i}]$ is an example of a transfer or chain matrix. It can be used to relate the potential coefficients of one layer to those of subsequent ones. For example, the coefficients of the $i+2$ th layer in terms of those of the i th layer are given by

$$\begin{bmatrix} A_{i+2,n} \\ B_{i+2,n} \\ C_{i+2,n} \\ D_{i+2,n} \end{bmatrix} = [T_{i+2,i+1}][T_{i+1,i}] \begin{bmatrix} A_{i,n} \\ B_{i,n} \\ C_{i,n} \\ D_{i,n} \end{bmatrix} \qquad (6.10)$$

The above chaining procedure is applicable so long as the strips do not enter into successive layers.

As mentioned earlier, the reason for the introduction of the transfer matrix is to obtain the Green's function of a multilayer planar structure systematically. This is demonstrated below.

By means of the transfer matrix, we first relate the potential function coefficients of the t th and the $t+1$ th dielectric layers to those of the lowest $(i=1)$ and the highest $(i=l)$

CHAP. 6 SPECTRAL DOMAIN SOLUTION OF MULTILAYER ... 96

dielectric layers in the structure, Fig. 6.4. We thus obtain the following equations:

$$\begin{bmatrix} A_{t,n} \\ B_{t,n} \\ C_{t,n} \\ D_{t,n} \end{bmatrix} = [T_{t,t-1}] \cdots [T_{2,1}] \begin{bmatrix} A_{1,n} \\ B_{1,n} \\ C_{1,n} \\ D_{1,n} \end{bmatrix} \tag{6.11}$$

$$\begin{bmatrix} A_{t+1,n} \\ B_{t+1,n} \\ C_{t+1,n} \\ D_{t+1,n} \end{bmatrix} = [T_{t+2,t+1}]^{-1} \cdots [T_{l,l-1}]^{-1} \begin{bmatrix} A_{l,n} \\ B_{l,n} \\ C_{l,n} \\ D_{l,n} \end{bmatrix} \tag{6.12}$$

Now let us turn to Eq. (3.19). From this equation it is easy to infer that at the interface where the strips are deposited, the tangential components of the field can be written as follows:

$$\begin{bmatrix} -j\beta \tilde{E}_{z,t} \\ -j\beta \tilde{H}_{z,t} \\ \tilde{E}_{x,t} \\ \tilde{H}_{x,t} \end{bmatrix} = [\gamma_t u_t] \begin{bmatrix} A_{t,n} \\ B_{t,n} \\ C_{t,n} \\ D_{t,n} \end{bmatrix} e^{-j\beta z} \tag{6.13}$$

$$\begin{bmatrix} -j\beta \tilde{E}_{z,t+1} \\ -j\beta \tilde{H}_{z,t+1} \\ \tilde{E}_{x,t+1} \\ \tilde{H}_{x,t+1} \end{bmatrix} = [\gamma_{t+1} u_{t+1}] \begin{bmatrix} A_{t+1,n} \\ B_{t+1,n} \\ C_{t+1,n} \\ D_{t+1,n} \end{bmatrix} e^{-j\beta z} \tag{6.14}$$

CHAP. 6   SPECTRAL DOMAIN SOLUTION OF MULTILAYER ...

Where $[\gamma_t u_t]$ is given by Eq. (6.7) with its v_i replaced by $u_t = h_t$ and $[\gamma_{t+1} u_{t+1}]$ is given by Eq. (6.8) with its u_i exchanged with $u_{t+1} = h_{t+1} - h_t$.

The last two equations can be related using the Fourier transforms of the appropriate boundary conditions at $y = h_t$. With reference to Eqs. (3.23), these boundary conditions can be written as follows:

$$\tilde{E}_{z,t} - \tilde{E}_{z,t+1} = 0 \tag{6.15.a}$$

$$\tilde{H}_{z,t} - \tilde{H}_{z,t+1} = -\tilde{J}_x e^{-j\beta z} \tag{6.15.b}$$

$$\tilde{E}_{x,t} - \tilde{E}_{x,t+1} = 0 \tag{6.15.c}$$

$$\tilde{H}_{x,t} - \tilde{H}_{x,t+1} = \tilde{J}_z e^{-j\beta z} \tag{6.15.d}$$

Using Eq. (6.11) and Eq. (6.12) in Eq. (6.13) and Eq. (6.14) respectively and substituting the results in Eqs. (6.15) lead to the following equation:

$$[K]\begin{bmatrix} A_{1,n} \\ B_{1,n} \\ C_{1,n} \\ D_{1,n} \end{bmatrix} - [L]\begin{bmatrix} A_{l,n} \\ B_{l,n} \\ C_{l,n} \\ D_{l,n} \end{bmatrix} = \begin{bmatrix} 0 \\ j\beta \tilde{J}_x \\ 0 \\ \tilde{J}_z \end{bmatrix} \tag{6.16.a}$$

where

$$[K] = [\gamma_t u_t][T_{t,t-1}] \cdots [T_{2,1}] \tag{6.16.b}$$

$$[L] = [\gamma_{t+1} u_{t+1}][T_{t+2,t+1}]^{-1} \cdots [T_{l,l-1}]^{-1} \tag{6.16.c}$$

From Eq. (6.16.a), it is clear that the potential function coefficients associated with the highest and lowest dielectrics in the structure are related. Since the shield is assumed to be a perfect conductor, we have the coefficients

$$B_{1,n} = C_{1,n} = B_{l,n} = C_{l,n} = 0 \tag{6.17}$$

This can be easily verified by examining Eqs. (6.2). Thus Eq. (6.16.a) can be written in a new form:

$$[K]\begin{bmatrix} A_{1,n} \\ 0 \\ 0 \\ D_{1,n} \end{bmatrix} - [L]\begin{bmatrix} A_{l,n} \\ 0 \\ 0 \\ D_{l,n} \end{bmatrix} = \begin{bmatrix} 0 \\ j\beta \tilde{J}_x \\ 0 \\ \tilde{J}_z \end{bmatrix} \qquad (6.18)$$

The above equation can be rearranged to take the following form:

$$\begin{bmatrix} A_{1,n} \\ D_{1,n} \\ A_{l,n} \\ D_{l,n} \end{bmatrix} = [V] \begin{bmatrix} 0 \\ j\beta \tilde{J}_x \\ 0 \\ \tilde{J}_z \end{bmatrix} \qquad (6.19)$$

where

$$[V] = [U]^{-1} \qquad (6.20)$$

Elements of the 4×4 matrix [U] are identical to those appearing in columns 1,4,5 and 8 of the 4×8 matrix [W] given by

$$[W] = [[K], -[L]] \qquad (6.21)$$

Equation (6.19) is, in fact, the key to finding the spectral domain Green's function of the problem. This is because it gives $A_{1,n}$ and $D_{1,n}$ in terms of the Fourier transforms of the components of the current distributions on the strips. When these $A_{1,n}$ and $D_{1,n}$ together with $B_{1,n}$ and $C_{1,n}$ given by Eq. (6.17), are substituted in Eq. (6.11) and the resulting equation is inserted in Eq. (6.13), we obtain the following equation

$$\begin{bmatrix} G_{11}(\alpha_n,\beta) & G_{12}(\alpha_n,\beta) \\ G_{21}(\alpha_n,\beta) & G_{22}(\alpha_n,\beta) \end{bmatrix} \begin{bmatrix} \tilde{J}_x \\ \tilde{J}_z \end{bmatrix} = \begin{bmatrix} \tilde{E}_z \\ \tilde{E}_x \end{bmatrix} \qquad (6.22)$$

where the right-hand side column matrix contains the Fourier transforms of the tangential electric field components at the interface where the strips are laid (for further detail, see Eqs. (3.38)) and

$$G_{11}(\alpha_n,\beta) = -(k_{11}v_{12}+k_{14}v_{22}) \qquad (6.23.a)$$

$$G_{12}(\alpha_n,\beta) = \frac{j}{\beta}(k_{11}v_{14}+k_{14}v_{24}) \qquad (6.23.b)$$

$$G_{21}(\alpha_n,\beta) = j\beta(k_{31}v_{12}+k_{34}v_{22}) \qquad (6.23.c)$$

$$G_{22}(\alpha_n,\beta) = (k_{31}v_{14}+k_{34}v_{24}) \qquad (6.23.d)$$

In the above expressions, k_{11}, k_{12} etc. are the elements of matrix $[K]$ given by Eq. (6.16.b) and v_{11}, v_{12} etc. are the elements of matrix $[V]$ given by Eq. (6.20).

Comparing Eq. (6.22) with Eq. (3.46) derived for the microstrip line, it is clear that in both cases, the orders of the matrices representing the spectral domain Green's functions are the same. This means that the number of dielectric layers does not have an effect on the order of the matrix specifying the Green's function in the spectral domain.

One important advantage of generating the Green's functions of a multilayer planar structure using the transfer matrix method is that the computing time would not increase more than linearly with the number of layers. By contrast, the method described by Farrar et al [11] would give computing time proportional to l^3 where l is the total number of dielectric layers. It should be mentioned that Farrar et al [11] have based their solution on the quasi-TEM approximation. That method is not going to be discussed here and it is left for the interested readers to examine it by themselves. It is believed that in view of materials presented in this chapter and Chapters 3 and 5, the reader should not have any difficulty in unravelling that work.

Now let us return to Eq. (6.16.a) and examine the case of a planar structure for which the top and/or bottom boundary is a magnetic wall. For instance, assume that the bottom wall is a magnetic wall. In this case, Eq. (6.17) reads as follows:

$$A_{l,n} = D_{l,n} = B_{l,n} = C_{l,n} = 0 \qquad (6.24)$$

Substituting the above conditions in Eq. (6.16.a) and following the same procedure as that already explained for obtaining Eq. (6.22), lead to the determination of the spectral domain Green's function of the problem of concern. This Green's function finds applications in the analysis of planar transmission lines which are symmetrical about the xz plane, Fig. 6.5. For instance, analysis of balanced microstrip lines

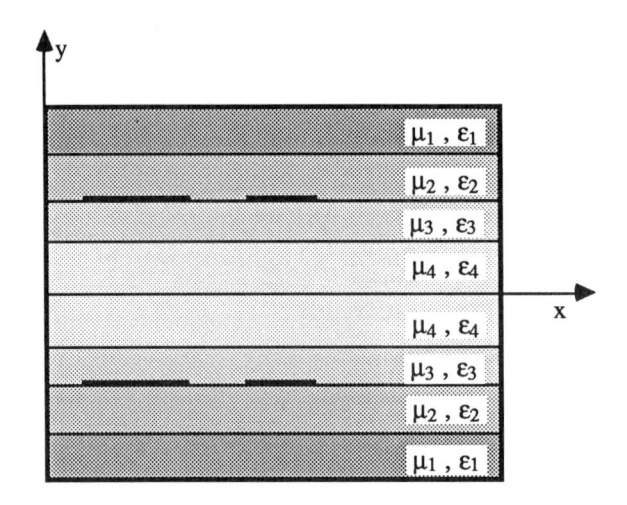

[Figure: Cross-section diagram showing layered structure with labels from top to bottom: μ₁,ε₁ ; μ₂,ε₂ ; μ₃,ε₃ ; μ₄,ε₄ ; μ₄,ε₄ ; μ₃,ε₃ ; μ₂,ε₂ ; μ₁,ε₁ — with x and y axes indicated.]

Fig. 6.5 Cross-section of a shielded multilayer multiconductor structure with symmetry about the xz plane.

comes under this category [12]. Note that if the existing symmetry is not considered in the solution of planar structures of the type shown in Fig. 6.5, then the resulting Green's functions will be 4×4 matrices instead of 2×2, in which case the solution will be achieved at a greater computing cost.

In view of Eq. (3.45) and other similar equations given in previous chapters, Eq. (6.22) can be regarded as the final equation in the spectral domain formulation of the multilayer planar structure shown in Fig. 6.4. By expanding currents or fields or a combination of these, Eq. (6.22) can be turned into a set of linear equations whose solutions yield the required unknowns. This is already detailed in Chapters 3, 4 and 5 and hence is redundant to be repeated here again.

6.2 Three-Layer Planar Transmission Lines with Coplanar Conductors

The cross-section of a three-layer planar transmission line is shown in Fig. 6.6. This structure can be viewed as a special case of the structure shown in Fig. 6.4. Since there are many microwave integrated circuits employing transmission lines of the type shown in Fig. 6.6, it is useful to have an explicit expression for the spectral domain Green's function associated with these lines. This Green's function which has been derived using the formulation in section 6.1, is given in this section. Prior knowledge

CHAP. 6 SPECTRAL DOMAIN SOLUTION OF MULTILAYER ...

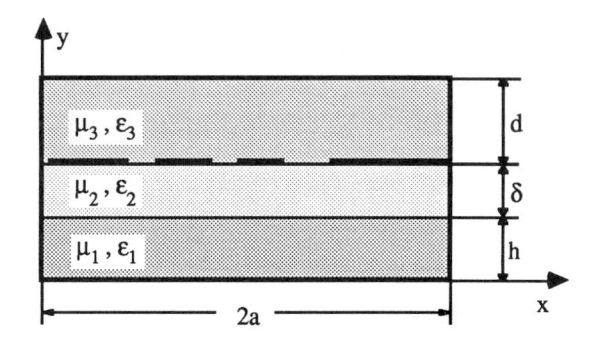

Fig. 6.6 Cross-section of a three-layer planar transmission line.

of the Green's function obviously reduces the burden of mathematics in achieving the spectral domain solution of Fig. 6.6. Also, in Appendix III, there is a computer program (subroutine) in Fortran which generates this Green's function when the third dielectric layer in Fig. 6.6 is air.

With reference to notations used in Eq. (6.22), the elements of the spectral domain Green's function of the structure shown in Fig. 6.6 are as follows:

$$G_{11}(\alpha_n,\beta) = -j\frac{k_3^2-\beta^2}{\beta}a_1 sinh(\gamma_{3,n}d) \qquad (6.25.a)$$

$$G_{12}(\alpha_n,\beta) = j\frac{k_3^2-\beta^2}{\beta}b_1 sinh(\gamma_{3,n}d) \qquad (6.25.b)$$

$$G_{21}(\alpha_n,\beta) = j(\alpha_n a_1 + \frac{\omega\mu_3\gamma_{3,n}}{\beta}a_2)sinh(\gamma_{3,n}d) \qquad (6.25.c)$$

$$G_{22}(\alpha_n,\beta) = -j(\alpha_n b_1 + \frac{\omega\mu_3\gamma_{3,n}}{\beta}b_2)sinh(\gamma_{3,n}d) \qquad (6.25.d)$$

In the above expressions:

$$a_1 = (d_o a''h'' - h_o f''a'' + d_o b''g'' + h_o l''b'')/\Delta$$

$$b_1 = (b_o a''f'' - d_o a''d'' - b_o l''b'' - d_o b''c'')/\Delta$$

$$a_2 = (a_o l''h'' + a_o f''g'' + l_o l''b'' - c_o a''h'' - c_o g''b'' - l_o a''f'')/\Delta$$

$$b_2 = (c_o a''d'' + c_o c''b'' - a_o l''d'' - a_o c''f'')/\Delta$$
$$\Delta = a''c_o(h_o d'' - h''b_o) + a''l_o(d''d_o - f''b_o) + b''c_o(c''h_o - b_o g'') +$$
$$b''l_o(l''b_o + d_o c'') + a_o c''(h''d_o - h_o f'') + a_o g''(f''b_o - d_o d'') +$$
$$a_o l''(h''b_o - d''h_o)$$

where

$$d'' = \frac{1}{\beta}\left[(k_1^2 - \beta^2)\cosh(\gamma_{2,n}\delta)\sinh(\gamma_{1,n}h) + \frac{\varepsilon_1 \gamma_{1,n}}{\varepsilon_2 \gamma_{2,n}}(k_2^2 - \beta^2)\sinh(\gamma_{2,n}\delta)\cosh(\gamma_{1,n}h)\right]$$

$$b'' = \frac{\alpha_n}{\omega \varepsilon_2 \gamma_{2,n}}(k_1^2 - k_2^2)\sinh(\gamma_{2,n}\delta)\cosh(\gamma_{1,n}h)$$

$$c'' = -\frac{\alpha_n}{\omega \gamma_{2,n}\mu_2}\left[(k_1^2 - k_2^2)\sinh(\gamma_{2,n}\delta)\sinh(\gamma_{1,n}h)\right]$$

$$d'' = \frac{1}{\beta}\left[(k_1^2 - \beta^2)\cosh(\gamma_{2,n}\delta)\cosh(\gamma_{1,n}h) + \frac{\mu_1 \gamma_{1,n}}{\mu_2 \gamma_{2,n}}(k_2^2 - \beta^2)\sinh(\gamma_{2,n}\delta)\sinh(\gamma_{1,n}h)\right]$$

$$l'' = -\alpha_n\left[\cosh(\gamma_{2,n}\delta)\sinh(\gamma_{1,n}h) + \frac{\varepsilon_1 \gamma_{1,n}}{\varepsilon_2 \gamma_{2,n}}\sinh(\gamma_{2,n}\delta)\cosh(\gamma_{1,n}h)\right]$$

$$g'' = \frac{1}{\beta \omega \mu_2 \gamma_{2,n}}\left[(k_1^2 k_2^2 - \beta^2 k_2^2 - \alpha_n k_1^2)\sinh(\gamma_{2,n}\delta)\sinh(\gamma_{1,n}h) - k_2^2 \gamma_{2,n}^2 \frac{\varepsilon_1 \gamma_{1,n}}{\varepsilon_2 \gamma_{2,n}} \times \right.$$
$$\left. \cosh(\gamma_{2,n}\delta)\cosh(\gamma_{1,n}h)\right]$$

$$h'' = -\alpha_n\left[\cosh(\gamma_{2,n}\delta)\cosh(\gamma_{1,n}h) + \frac{\mu_1 \gamma_{1,n}}{\mu_2 \gamma_{2,n}}\sinh(\gamma_{2,n}\delta)\sinh(\gamma_{1,n}h)\right]$$

$$a_o = \frac{k_3^2 - \beta^2}{\beta}\sinh(\gamma_{3,n}d)$$

$$b_o = \frac{k_3^2 - \beta^2}{\beta} \cosh(\gamma_{3,n} d)$$

$$c_o = -\alpha_n \sinh(\gamma_{3,n} d)$$

$$d_o = -\frac{\omega \mu_3 \gamma_{3,n}}{\beta} \sinh(\gamma_{3,n} d)$$

$$l_o = \frac{\omega \varepsilon_3 \gamma_{3,n}}{\beta} \cosh(\gamma_{3,n} d)$$

$$h_o = -\alpha_n \cosh(\gamma_{3,n} d)$$

For structures with an arbitrary distribution of conductors, Fig. 6.6, the spectrum parameter α_n appearing in the above equations is given by:

$$\alpha_n = \frac{n\pi}{2a} \tag{6.26.a}$$

However, if conductors are symmetrically distributed about the yz plane, this parameter is given by:

$$\alpha_n = \frac{n\pi}{a} \tag{6.26.b}$$

when an electric wall is assumed at the symmetry plane, Fig. 6.7.a, and by

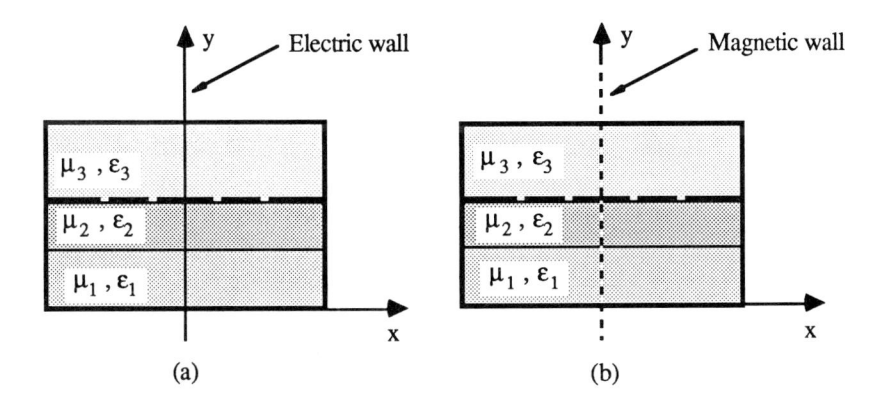

Fig. 6.7 Cross-section of a shielded three-layer planar transmission line with symmetry about the yz plane; (a) an electric wall at the symmetry plane and (b) a magnetic wall at the symmetry plane.

$$\alpha_n = (n+\tfrac{1}{2})\frac{\pi}{a} \qquad (6.26.c)$$

when a magnetic wall is imagined at the yz plane, Fig. 6.7.b. The reason for the above classification is already presented in Chapter 3 and section 5.2.1.

Before proceeding to a further generalization of the spectral domain technique, it would be appropriate at this stage to briefly discuss the spectral domain solutions of some widely used three-layer planar transmission lines.

6.2.1 Coplanar Waveguides

The cross-section of a coplanar waveguide is shown in Fig. 6.8.a. This structure is originally introduced by Wen [13], who also presents its quasi-TEM solution using conformal mapping. In order to simplify the analysis, Wen assumes that the dielectric layer in the coplanar waveguide is infinitely thick. This assumption, however, introduces some errors in the evaluation of characteristics of coplanar waveguides whose slots sizes are of the same order or larger than the thickness of the substrate. Based on the quasi-static approximation, Davis et al [14] later present a more accurate solution of the coplanar waveguide, taking into account the effect of substrate thickness. However, because of the quasi-TEM nature of both solutions [13,14], their methods are incapable of predicting the dispersion characteristics of the coplanar waveguides.

With reference to Fig. 6.8.a, it is clear that the coplanar waveguide is a special case

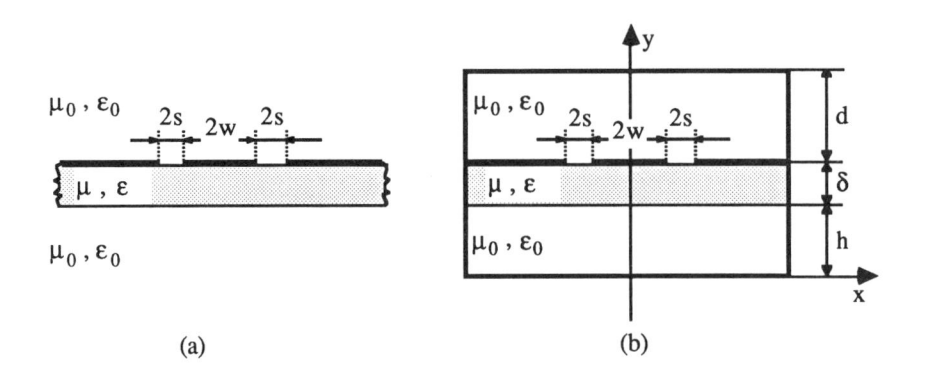

Fig. 6.8 Cross-section of (a) an open coplanar waveguide and (b) a shielded coplanar waveguide.

CHAP. 6 SPECTRAL DOMAIN SOLUTION OF MULTILAYER ... 105

of the three-layer planar structure shown in Fig. 6.6. Thus, its spectral domain full-wave analysis can be carried out using the Green's function given by Eqs. (6.25). We may recall that in the derivation of the Green's function, we assumed that the structure is shielded. Hence, we can analyse the shielded version of the coplanar waveguide, Fig. 6.8.b, without any difficulty.

For the coplanar waveguide, the normal mode of operation is the even mode. With this mode, one can associate a magnetic wall located at the symmetry plane ; $x = 0$. Hence, in the analysis of this case, the spectral parameter α_n has to be chosen according to Eq. (6.26.c). The coplanar waveguide can also support odd modes, in which case an electric wall can be assumed at the plane of symmetry. Thus, its corresponding α_n is given by Eq. (6.26.b).

Results of the spectral domain analysis of a coplanar waveguide are shown in Fig. 6.9. As this figure shows, the fundamental mode (normal mode) of operation which is an even mode has no cut-off frequency. On the contrary, the odd mode has a cut-off frequency and as shown in the subsequent figure, Fig. 6.10, its value depends

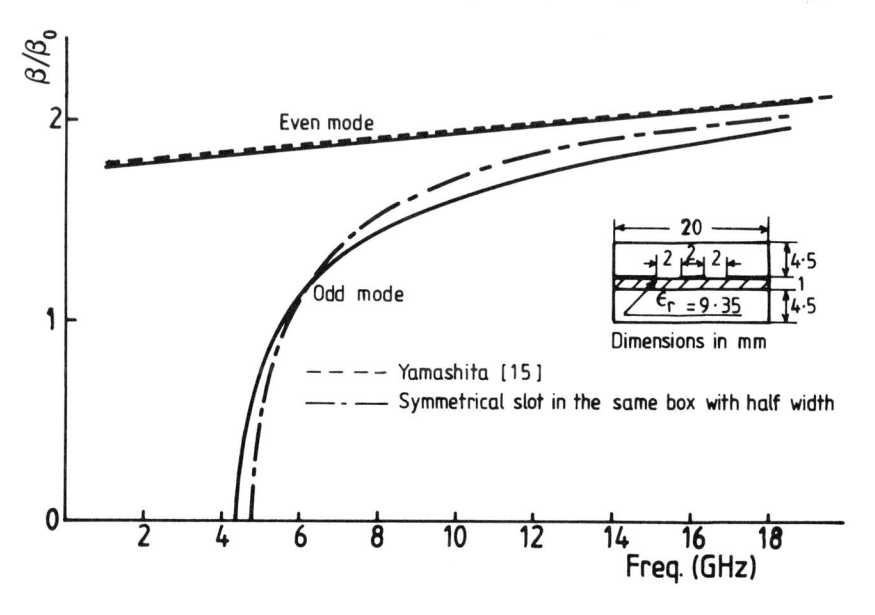

Fig. 6.9 Dispersion characteristics of the even and the odd mode of a shielded coplanar waveguide computed by the spectral domain technique. (Result shown in - - - has been computed by "nonuniform discretization of integral equations", from Yamashita and Atsuki [15], Copyright © 1976 IEEE, reproduced with permission.)

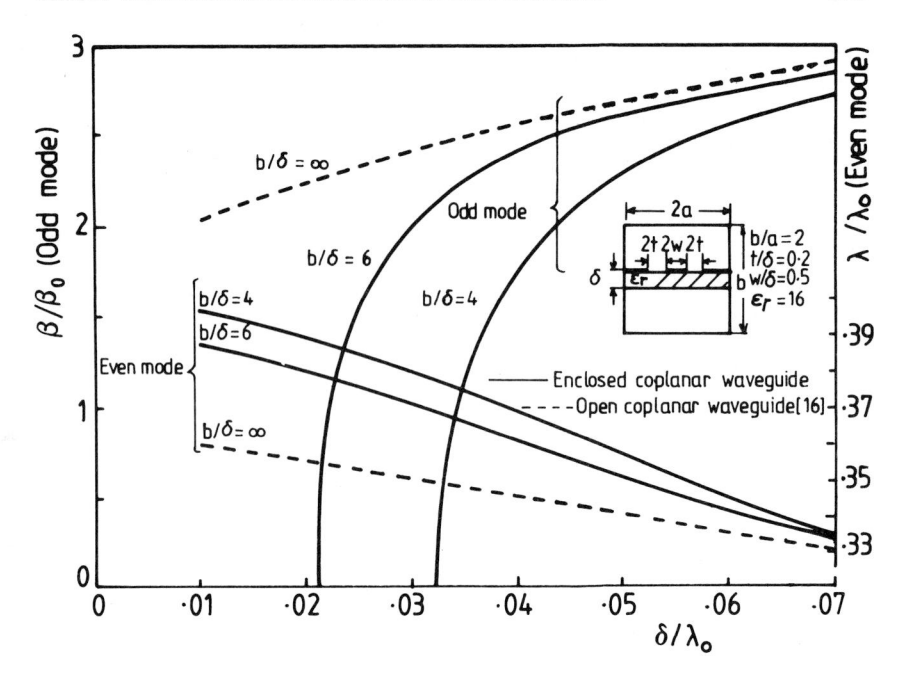

Fig. 6.10 Dispersion characteristics of shielded and open coplanar waveguides computed by the spectral domain technique. (Open coplanar waveguide results from Knorr and Kuckler [16], Copyright © 1975 IEEE, reproduced with permission.)

on the size of the shield. In Fig. 6.9 and Fig. 6.10, results of other investigators' work, Yamashita et al [15] and Knorr et al [16], are depicted for comparison.

In the analysis whose results are illustrated in Fig. 6.9 and Fig. 6.10, Eq. (4.22) instead of Eq. (6.22) was used. This is because the sizes of slots in the coplanar waveguide are usually much less than the sizes of strips and hence according to section 4.4, solution with Eq. (4.22) should converge faster. As far as the expansions of the field components $E_x(x)$ and $E_z(x)$ are concerned, the basis functions are chosen to be Legendre polynomials and sinusoidal functions. Thus, within the right-hand side slot, the required field components are expanded as follows:

$$E_x(x) = \sum_{q=1}^{Q} b_q P_{m-1}\left(\frac{x-w-s}{s}\right) \quad (6.27.a)$$
$$w < x < w+2s$$

$$E_z(x) = \sum_{p=1}^{P} a_p \sin\left[m\pi\left(\frac{x-w}{2s}\right)\right] \quad (6.27.b)$$

CHAP. 6 SPECTRAL DOMAIN SOLUTION OF MULTILAYER... 107

from which, depending on whether the even or the odd mode characteristics are required, the field components for the left slot can be readily formed. Further information as regards the above expansions as well as other basis functions are given in section 4.1. Dispersion curves in Fig. 6.9 and Fig. 6.10 are produced using $P=Q=5$ corresponding to an accuracy of better than 0.1% in the calculation of β. Note that with the above values for P and Q, the order of the final matrix is $P + Q = 10$.

6.2.2 Inverted Microstrip Lines

The cross-section of an inverted microstrip line is shown in Fig. 6.11.a. Schneider [17] and Spielman [18] apply the quasi-TEM approximation in order to characterise this transmission line. It is clear that at high frequencies, the quasi-TEM solution cannot be acceptable. This can be readily established by making an analogy between this line and the microstrip line.

Dispersion characteristics of the inverted microstrip line can be investigated via the spectral domain technique. Considering the shielded version of this line, Fig. 6.11.b, we already have its spectral domain Green's function, Eqs. (6.25). Therefore, in order to convert Eq. (6.22) into the final matrix equation, we need to expand $J_x(x)$ and $J_z(x)$ in terms of appropriate basis functions.

After choosing the basis functions identical to those given by Eq. (4.7), a set of curves showing the dispersive behaviour of the fundamental mode of a shielded inverted microstrip line was computed and is depicted in Fig. 6.12. Each curve

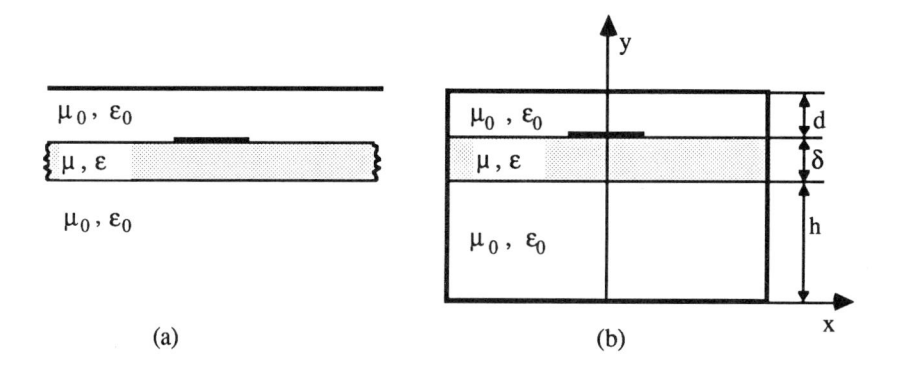

Fig. 6.11 Cross-section of (a) an open inverted microstrip line and (b) a shielded inverted microstrip line.

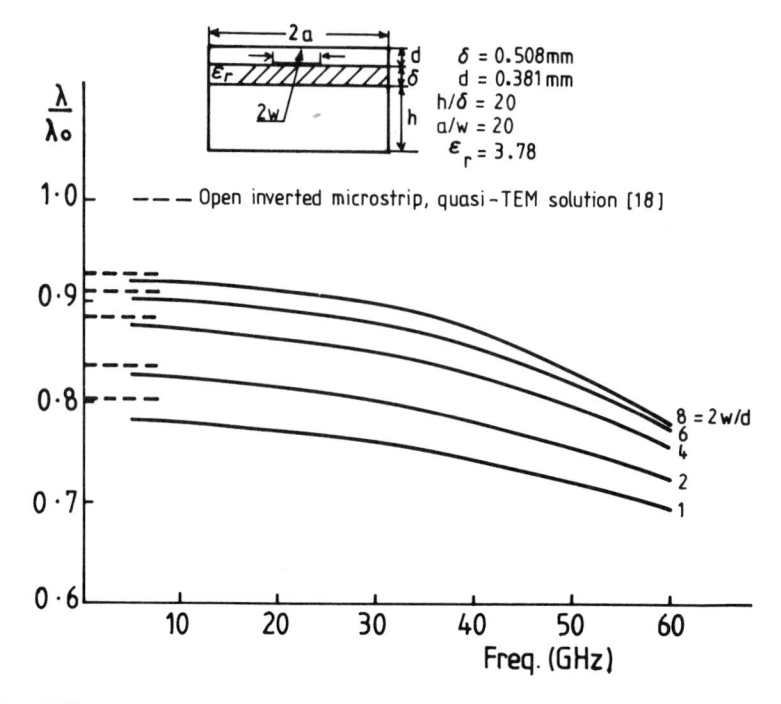

Fig. 6.12 Dispersion characteristics of the fundamental mode of a shielded inverted microstrip line for various strip sizes computed by the spectral domain technique. (Quasi-TEM results from Spielman [18], Copyright © 1977 IEEE, reproduced with permission.)

corresponds to a particular strip width. The presented dispersion curves were obtained using a three-term expansion for both $J_x(x)$ and $J_z(x)$ in which case the numerical accuracy in the calculation of β was set to be better than 0.1%. In Fig. 6.12, the quasi-TEM results reported by Spielman [18] are also shown. Note that Spielman's results are for the open version of the shielded inverted microstrip line whose dispersion characteristics have been computed.

6.2.3 Finlines

Finlines, originally developed from ridge waveguides [19], are widely used in the development of microwave integrated circuits at millimetre wavelengths [5,6,20-23]. Cross-sections of two practical finlines are shown in Fig. 6.13 with 6.13.a the more conventional form. In the manufacturing process of finlines, the substrate on which

CHAP. 6 SPECTRAL DOMAIN SOLUTION OF MULTILAYER... 109

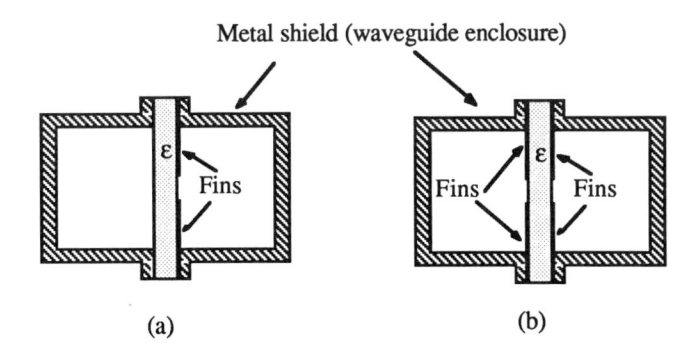

Fig. 6.13 Cross-section of (a) a unilateral finline and (b) a bilateral finline in practical configurations.

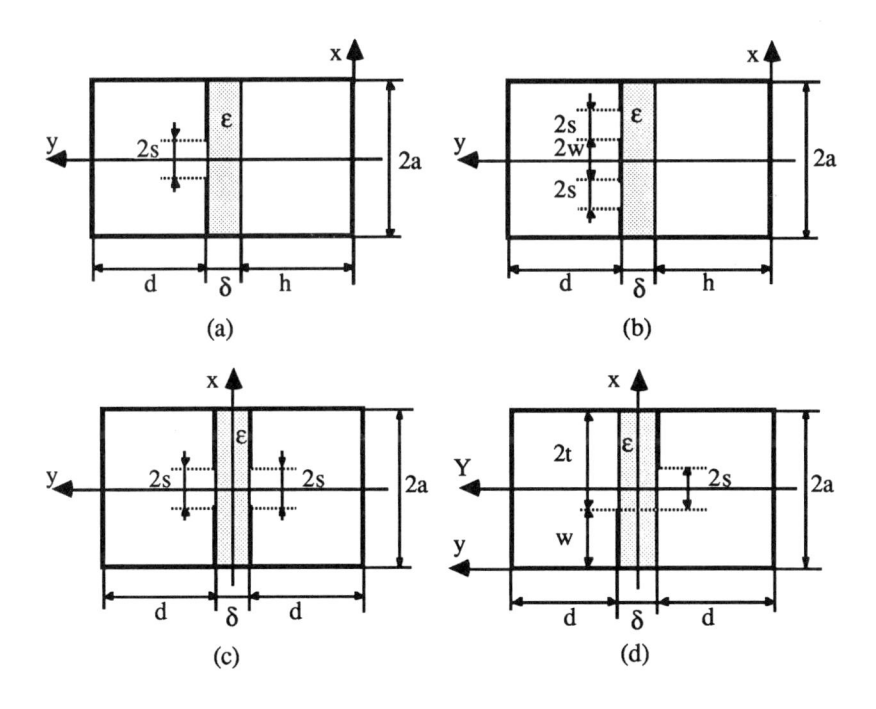

Fig. 6.14 Models usually used in the analysis of (a) unilateral finlines, (b) coupled slot finlines, (c) bilateral finlines and (d) antipodal finlines.

the finline metallization is etched, is clamped between the two halves of an appropriate waveguide. Since the electromagnetic energy is mostly confined to the gap between fins, no appreciable energy leak can be associated with the opening between the two parts of the finline enclosure. In view of this fact, one sensible model for the finline shield is a fully connected rectangle, Fig. 6.14. This model significantly reduces the mathematical complexity in analysing finlines.

The first two finlines whose spectral domain solutions are initially discussed, are the unilateral finline and the coupled slot finline, Fig. 6.14.a and Fig. 6.14.b. Green's functions associated with these finlines are already available and are given by Eqs. (6.25). In fact, the spectral domain solution of the coupled slot finline is exactly the same as that explained for the coplanar waveguide in section 6.2.1. Therefore, no further comment will be made in this regard, and the interested reader is referred to publication [24].

As far as the spectral domain solution of the unilateral finline is concerned, it is in principle the same as that explained in the case of the inverted microstrip line. In fact, this becomes clearly discernible if the slot and the strip are treated as dual elements. Therefore, we may use the same type of basis functions to expand the unknown field or current components in either case. In publication [24], the basis functions chosen for the analysis of the fundamental mode of the unilateral finline are:

$$E_{z,q}(x) = sin(q\pi X) \qquad (6.28.a)$$
$$E_{x,p}(x) = P_{2(p-1)}(X) \qquad 0<|x|<1 \qquad (6.28.b)$$

where $X = x/s$ and s is the half-width of the slot, Fig. 6.14.a. The above basis functions are in the forms introduced by Eqs. (4.7). They have been used to compute the dispersion characteristics of the shielded inverted microstrip line, see section 6.2.2. We recall that in order to form the final matrix, the above equations must be used in conjunction with Eq. (4.22) in which case the Green's function given by Eq. (6.25) has to be converted into a Green's function suitable for application in Eq. (4.22). The relation between these two Green's functions is given by Eq. (3.50).

It is worth pointing out that the similarity among solutions of the microstrip line, the inverted microstrip line, the slot line, the coupled slot line, the coplanar waveguide and the coupled microstrip line is so much that a computer program developed for one of these structures can be easily modified to work for the other ones as well. In Appendix III, there is a computer program which can analyse these structures.

As a numerical example, for a particular unilateral finline, variations of the

CHAP. 6 SPECTRAL DOMAIN SOLUTION OF MULTILAYER... 111

normalised wave-length with the slot width for three different frequencies have been computed and are shown in Fig. 6.15. These data, which are for the fundamental mode, have been generated using the basis functions given by Eqs. (6.28). In these computations, three basis functions for each field component were used. Thus, the order of the final matrix was 6. In order to reduce the matrix order and hence to speed up the computation process, the following approximation is applicable.

In finlines, the dielectric is very thin so that these structures can be simply assumed as a perturbation to ridge waveguides when the ridges are infinitely thin [25]. Since ridge waveguides support TE or TM modes, it is possible to assume that for the fundamental mode of the unilateral finline, $E_z(x)$ is negligible. This is due to the fact that this mode resembles the fundamental mode of double ridge waveguide which is TE [19]. Thus, by eliminating $E_z(x)$ from the spectral domain mathematics, the order of the final matrix is automatically reduced. Implementation of the above approximation in the above example reduces the matrix order to 3. As a result, computations can be carried out four times faster.

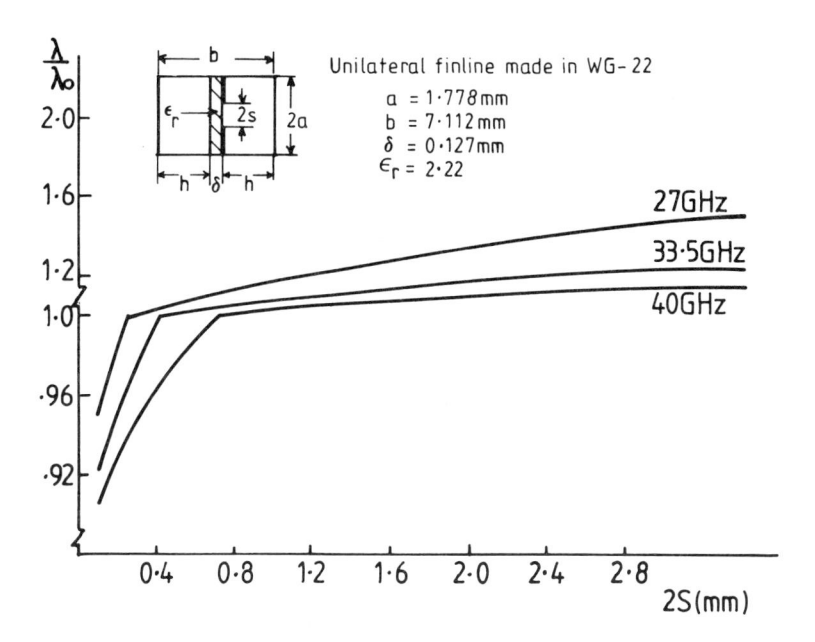

Fig. 6.15 Normalised wavelength of the fundamental mode of a unilateral finline against the slot size computed by the spectral domain technique for three different frequencies.

Knorr et al [26] in the spectral domain analysis of the unilateral finline, use a further approximation and assume that in the slot, $E_x(x)$ and $E_z(x)$ are

$$E_x(x) = 1 \qquad (6.29.a)$$
$$E_z(x) = 0 \qquad |x| \leq s \qquad (6.29.b)$$

The consequence of such approximation is already seen in respect of the solution of the microstrip line; see Eq. (4.14). This approximation reduces the determinant of the final matrix to

$$\sum_n H_{12}(\alpha_n,\beta) |\tilde{E}_x|^2 = 0 \qquad (6.30)$$

The above approximation, however, imposes some limitations on the accuracy of the solution, particularly, if the finline has a large gap. But if a few percent error can be tolerated, Eq. (6.30) would be ideal for CAD applications.

Now let us return to the two remaining finlines, the bilateral finline and the antipodal finline shown in Fig. 6.14, and discuss their solutions using the spectral domain technique. It is clear that due to the existence of two layers of metallization in these finlines, Eqs.(6.25) do not represent the Green's functions of these structures. Nevertheless, it can be exploited to generate the desired Green's functions.

We consider the bilateral finline first, Fig. 6.14.c. This structure enjoys a symmetry about the xz plane. When the two slots in the bilateral finline are excited in phase (even mode), a magnetic wall can be imagined at the symmetry plane. Therefore, the analysis of this structure can be carried out by considering half of it only, Fig. 6.16.a. Since at one side the new structure is enclosed by the magnetic wall, its associated Green's function can be obtained using conditions (6.24) in a manner explained in section 6.1. It is not difficult to arrive at the conclusion that the required Green's

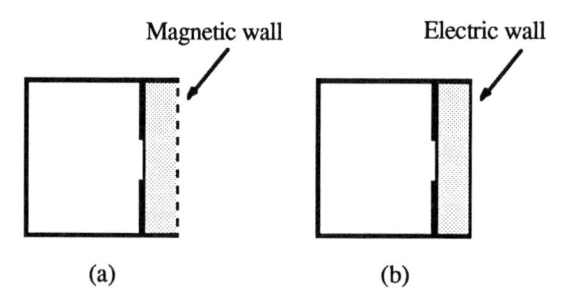

Fig. 6.16 Cross-section of the model used in the analysis of (a) the even mode and (b) the odd mode of the bilateral finline.

CHAP. 6 SPECTRAL DOMAIN SOLUTION OF MULTILAYER... 113

function can be formed by taking Eqs. (6.25) and replacing

$$sinh(\gamma_{l,n}h) \quad with \quad cosh(\gamma_{l,n}h) \tag{6.31}$$

and vice versa [24]. Furthermore, since there are only two layers in Fig. 6.16.a, δ in Eqs. (6.25) has to be equated to zero.

Having obtained the spectral domain Green's function of the bilateral finline by the above technique, one can then set up the final matrix equation by following exactly the same approach as that explained for the unilateral finline. It should be mentioned that the first computer program presented in Appendix III can be easily modified to a program capable of computing the dispersion characteristics of the bilateral finline. This modification has to be made only to the subroutine generating the Green's function (see Appendix III). For a particular bilateral finline, a set of normalised wavelengths against the slot width at three different frequencies has been computed and shown in Fig. 6.17. These results were generated with the computer program in Appendix III after the necessary changes had been made to it. In this analysis, the

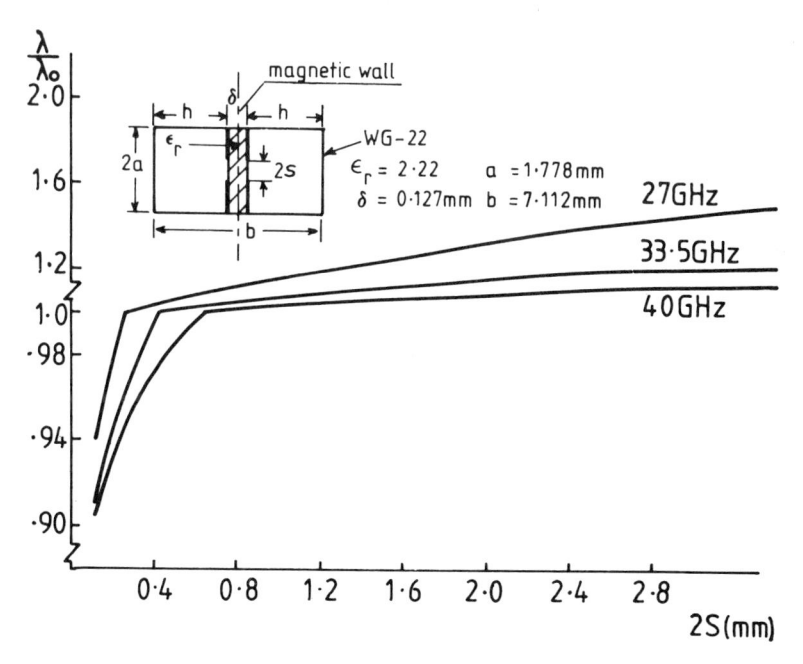

Fig. 6.17 Normalised wavelength of the fundamental mode of a bilateral finline against the slot size computed by the spectral domain technique for three different frequencies.

type and number of basis functions were kept the same as those used for the analysis of the unilateral finline shown in Fig. 6.15.

In the analysis of the bilateral finline, we may also consider the case that the two slots are excited antiphase (odd mode). Solution to this case is equivalent to that of solving one half of the bilateral finline structure with an electric wall at the symmetry plane [27,28], Fig. 6.16.b. It is clear that the emerged structure can be analysed using the unilateral finline Green's function, Eq. (6.25). Therefore, the first computer program in Appendix III can be used to analyse the odd mode of the bilateral finline.

Now let us examine how the Green's function of the antipodal finline, Fig. 6.14.d, can be derived using Eqs. (6.25). Unlike the bilateral finline, the antipodal finline is not symmetrical about the xz plane. But it enjoys a rotational symmetry about the Z axis; by rotating the structure through 180° about the Z-axis, one arrives at the same structure. In a manner described below, it is possible to take advantage of this symmetry so as to generate the antipodal finline Green's function from Eqs. (6.25). To illustrate this, we consider a more complicated antipodal finline shown in Fig. 6.18.

Like other multilayer structures studied so far, the spectral domain potential functions associated with each dielectric layer of the new antipodal finline, Fig. 6.18, are given by Eqs. (6.2). For the layer with index $i = 1$, these potential functions in the space domain are:

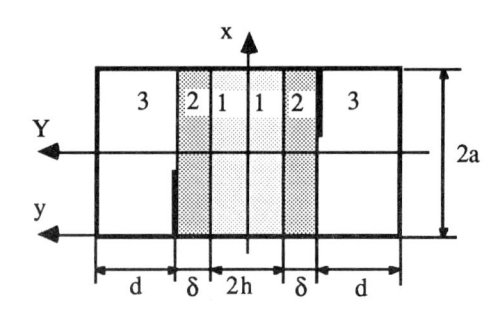

$$\phi_1 = \sum_n [A_{1,n}\sinh(\gamma_{1,n}y) + B_{1,n}\cosh(\gamma_{1,n}y)]\sin(\alpha_n x) \quad (6.32.a)$$

$$\psi_1 = \sum_n [C_{1,n}\sinh(\gamma_{1,n}y) + D_{1,n}\cosh(\gamma_{1,n}y)]\cos(\alpha_n x) \quad (6.32.b)$$

Fig. 6.18 Cross-section of a six-layer antipodal finline.

CHAP. 6 SPECTRAL DOMAIN SOLUTION OF MULTILAYER...

where α_n is chosen according to Eq. (6.26.a);

$$\alpha_n = \frac{n\pi}{2a} \qquad n = 0,1,2,... \qquad (6.33)$$

Since the structure is rotationally symmetric, the following conditions should hold at $y = 0$:

$$\phi_1(x,0) = \phi_1(2a-x,0) \qquad (6.34.\text{a})$$

$$\psi_1(x,0) = \psi_1(2a-x,0) \qquad (6.34.\text{b})$$

The above conditions impose a constraint on the values of $B_{1,n}$ and $D_{1,n}$, namely,

$$B_{1,n} = 0 \qquad \text{for} \quad n = 0,2,4,... \qquad (6.35.\text{a})$$

$$D_{1,n} = 0 \qquad \text{for} \quad n = 1,3,5,... \qquad (6.35.\text{b})$$

The rotational symmetry also requires that the above potential functions satisfy the following conditions at $x = a$

$$\phi_1(a,y) = \phi_1(a,-y) \qquad (6.36.\text{a})$$

$$\psi_1(a,y) = \psi_1(a,-y) \qquad (6.36.\text{b})$$

These conditions are met only if the potential coefficients $A_{1,n}$ and $C_{1,n}$ are

$$A_{1,n} = 0 \qquad \text{for} \quad n = 1,3,5,... \qquad (6.37.\text{a})$$

$$C_{1,n} = 0 \qquad \text{for} \quad n = 0,2,4,... \qquad (6.37.\text{b})$$

Considering conditions (6.35) and (6.37), it is easy to conclude that as n takes $0,1,2,...$ the potential coefficients in Eqs. (6.32) alternate between

$$\begin{bmatrix} A_{1,n} \\ 0 \\ 0 \\ D_{1,n} \end{bmatrix} \quad \text{and} \quad \begin{bmatrix} 0 \\ B_{1,n} \\ C_{1,n} \\ 0 \end{bmatrix} \qquad (6.38)$$

The above conditions correspond to Eq. (6.24) and Eq. (6.17) representing the boundary conditions for the magnetic and electric wall at the bottom of the three-layer structure, Fig. 6.6. Therefore, as n takes $0,1,2,...$, the spectral domain Green's function of the antipodal finline can be assumed to alternate between the Green's function of the bilateral finline and the unilateral finline. The method of generating

CHAP. 6   SPECTRAL DOMAIN SOLUTION OF MULTILAYER...   116

these Green's functions using Eqs. (6.25) is already discussed in this section. Thus the Green's function of the antipodal finline is theoretically available.

The above technique has been used in [24] to calculate the dispersion characteristics of the antipodal finline. In that work where Eq. (4.22) has been used, $E_x(x)$ and $E_z(x)$ are expanded as follows:

$$E_x(x) = \sum_{q=1}^{Q} b_q P_{q-1}\left(\frac{x-w-t}{t}\right) \tag{6.39.a}$$

$$E_z(x) = \sum_{p=1}^{P} a_p \sin\left[\frac{p\pi}{2t}(x-w)\right] \tag{6.39.b}$$

where w and t are shown in Fig. 6.14.d. Results of the computation of the normalised wavelength of an antipodal finline are shown in Fig. 6.19. These results are achieved using the above basis functions after truncating the series at four terms $(P = Q = 4)$. In this case, the error in computation of β was set to be smaller than 0.1%.

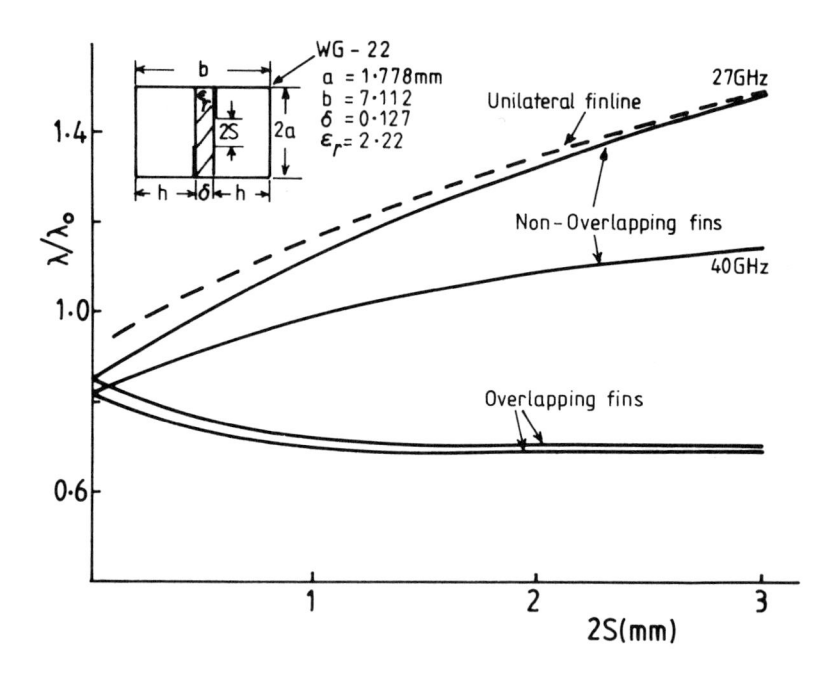

Fig. 6.19 Normalised wavelength of the fundamental mode of an antipodal finline against fins separation computed by the spectral domain technique for two different frequencies. (From Mirshekar-Syahkal and Davies [24], Copyright © 1982 IEEE, reproduced with permission.)

6.3 Spectral Domain Solution of Multilayer Transmission Lines with Multilayer Conductors

In this section we discuss the application of the spectral domain technique in the analysis of multilayer transmission lines with conductors arranged at different dielectric interfaces, Fig. 6.20. In this application, the problem is formulated using the transfer matrix approach introduced in section 6.1.

With reference to Fig. 6.20, coefficients of the potential functions of layers $t,u,v,w,...$ can be related to those of layers $1, t+1, u+1, v+1,...$ respectively using Eq. (6.5). The resulting expressions are as follows:

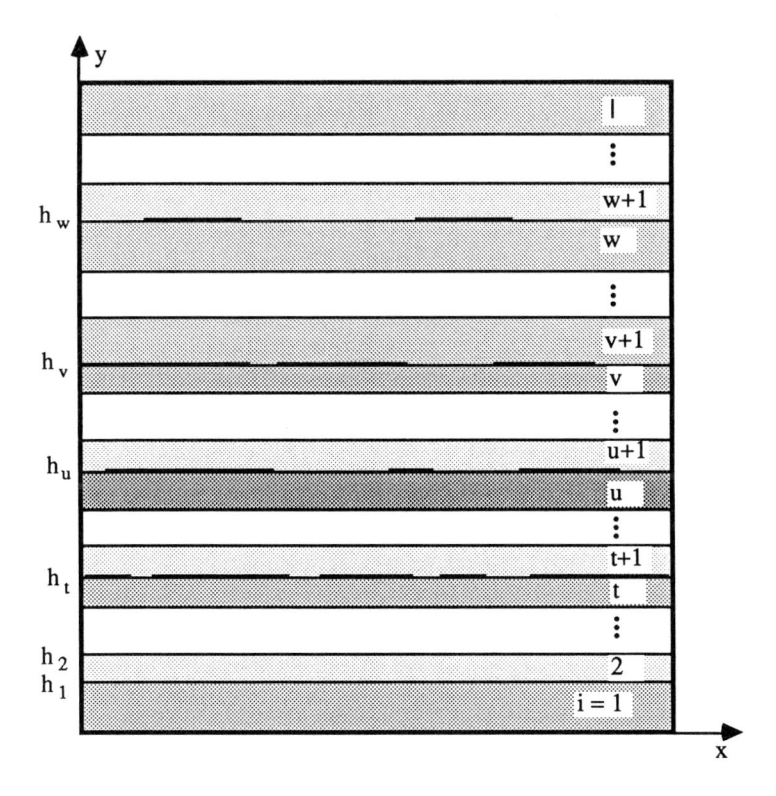

Fig. 6.20

$$\begin{bmatrix} A_{t,n} \\ B_{t,n} \\ C_{t,n} \\ D_{t,n} \end{bmatrix} = \begin{bmatrix} T_{t,t-1} \end{bmatrix} \cdots \begin{bmatrix} T_{2,1} \end{bmatrix} \begin{bmatrix} A_{1,n} \\ B_{1,n} \\ C_{1,n} \\ D_{1,n} \end{bmatrix} \qquad (6.40.a)$$

$$\begin{bmatrix} A_{u,n} \\ B_{u,n} \\ C_{u,n} \\ D_{u,n} \end{bmatrix} = \begin{bmatrix} T_{u,u-1} \end{bmatrix} \cdots \begin{bmatrix} T_{t+2,t+1} \end{bmatrix} \begin{bmatrix} A_{t+1,n} \\ B_{t+1,n} \\ C_{t+1,n} \\ D_{t+1,n} \end{bmatrix} \qquad (6.40.b)$$

$$\begin{bmatrix} A_{v,n} \\ B_{v,n} \\ C_{v,n} \\ D_{v,n} \end{bmatrix} = \begin{bmatrix} T_{v,v-1} \end{bmatrix} \cdots \begin{bmatrix} T_{u+2,u+1} \end{bmatrix} \begin{bmatrix} A_{u+1,n} \\ B_{u+1,n} \\ C_{u+1,n} \\ D_{u+1,n} \end{bmatrix} \qquad (6.40.c)$$

$$\begin{bmatrix} A_{w,n} \\ B_{w,n} \\ C_{w,n} \\ D_{w,n} \end{bmatrix} = \begin{bmatrix} T_{w,w-1} \end{bmatrix} \cdots \begin{bmatrix} T_{v+2,v+1} \end{bmatrix} \begin{bmatrix} A_{v+1,n} \\ B_{v+1,n} \\ C_{v+1,n} \\ D_{v+1,n} \end{bmatrix} \qquad (6.40.d)$$

$$\vdots$$

It is clear that the above equations are similar to Eqs. (6.11) and (6.12). Therefore, boundary conditions at the interfaces of layers t and $t+1$, u and $u+1$, v and $v+1$, w and $w+1$,... can be utilised in order to relate the above equations. These boundary conditions are essentially the same as those in Eqs. (6.15) which in terms of the potential functions coefficients (see Eqs. (6.13) and (6.14)) are given as follows:

$$[\gamma_t u_t]\begin{bmatrix}A_{t,n}\\B_{t,n}\\C_{t,n}\\D_{t,n}\end{bmatrix} - [\gamma_{t+1} u_{t+1}]\begin{bmatrix}A_{t+1,n}\\B_{t+1,n}\\C_{t+1,n}\\D_{t+1,n}\end{bmatrix} = \begin{bmatrix}0\\j\beta\tilde{J}_{x,t}\\0\\\tilde{J}_{z,t}\end{bmatrix} \qquad (6.41.\text{a})$$

$$[\gamma_u u_u]\begin{bmatrix}A_{u,n}\\B_{u,n}\\C_{u,n}\\D_{u,n}\end{bmatrix} - [\gamma_{u+1} u_{u+1}]\begin{bmatrix}A_{u+1,n}\\B_{u+1,n}\\C_{u+1,n}\\D_{u+1,n}\end{bmatrix} = \begin{bmatrix}0\\j\beta\tilde{J}_{x,u}\\0\\\tilde{J}_{z,u}\end{bmatrix} \qquad (6.41.\text{b})$$

$$[\gamma_v u_v]\begin{bmatrix}A_{v,n}\\B_{v,n}\\C_{v,n}\\D_{v,n}\end{bmatrix} - [\gamma_{v+1} u_{v+1}]\begin{bmatrix}A_{v+1,n}\\B_{v+1,n}\\C_{v+1,n}\\D_{v+1,n}\end{bmatrix} = \begin{bmatrix}0\\j\beta\tilde{J}_{x,v}\\0\\\tilde{J}_{z,v}\end{bmatrix} \qquad (6.41.\text{c})$$

$\vdots$

In the above relations, subscripts $t,u,v,\ldots$ assigned to the current components, specify the interfaces between dielectric layers t and $t+1$, u and $u+1$,... respectively. Note that at these interfaces, conductors are present. Also note that in the above equations, matrices $[\gamma_t u_t]$, $[\gamma_{t+1} u_{t+1}]$ etc. are similar to those in Eqs. (6.13) and (6.14). Substituting Eqs. (6.40) in Eqs. (6.41) leads to a matrix expression relating the coefficients of potential functions of the first and last layers to the Fourier transforms of the components of currents on strips. For multilayer transmission lines with coplanar conductors, this matrix expression is indeed Eq. (6.16.a). In view of this, the matrix expression associated with the general case of multilayer multiconductor structure, Fig. 6.20, can be reduced to a form similar to Eq. (6.19) when the boundary conditions of the top and bottom walls are applied. The resulting equation is then used in a similar manner explained in section 6.1 in order to set up the final equations in the spectral domain and to obtain the spectral domain Green's function of the problem. In matrix notation, the final equations appear as follows:

$$\begin{bmatrix} G_{11}(\alpha_n,\beta) & G_{12}(\alpha_n,\beta) & \cdots \\ G_{21}(\alpha_n,\beta) & G_{22}(\alpha_n,\beta) & \cdots \\ G_{31}(\alpha_n,\beta) & & \cdots \\ \cdots & & \\ \cdots & & \end{bmatrix} \begin{bmatrix} \tilde{J}_{x,t} \\ \tilde{J}_{z,t} \\ \tilde{J}_{x,u} \\ \tilde{J}_{z,u} \\ \vdots \end{bmatrix} = \begin{bmatrix} \tilde{E}_{x,t} \\ \tilde{E}_{z,t} \\ \tilde{E}_{x,u} \\ \tilde{E}_{z,u} \\ \vdots \end{bmatrix} \quad (6.42)$$

Comparing Eq. (6.42) with Eq. (3.45), it is clear that the left-hand side matrix in Eq. (6.42) represents the spectral domain Green's function of the multilayer multiconductor transmission line, Fig. 6.20. It is easy to conclude that the order of this matrix depends on the number of conductor planes involved in the structure. In fact, for a structure with l_s conductor layers, the order of this matrix is $2l_s$. In Eq. (6.42), the right-hand side column matrix contains the Fourier transforms of the tangential electric field components associated with interfaces where strips exist. Parameters $G_{11}(\alpha_n,\beta)$, $G_{12}(\alpha_n,\beta)$ etc. appearing in Eq. (6.42) are not easily expressible in explicit forms. However, for some specific cases, these are analytically available [29-31].

A solution to Eq. (6.42) can be obtained using the moment method developed in Chapter 3. Initially $J_{x,t}(x)$, $J_{z,t}(x)$, $J_{x,u}(x)$, ... are expressed in series of suitable basis functions. The Galerkin technique is then applied in the Fourier domain, converting Eq. (6.42) into a set of homogeneous equations. From these equations, the phase constant β can be computed in a way similar to that explained in Chapter 3 for the microstrip line. As pointed out in section 4.4, in setting up the characteristic (final) matrix, we are not necessarily limited to the use of Eq. (6.42). Other versions of the same equation can also be useful. In fact, depending on the widths of slots and strips in a structure of concern, Eq. (6.42) should be rearranged so that unknown functions (currents and/or fields) can be efficiently approximated. The following example clarifies this matter.

CHAP. 6 SPECTRAL DOMAIN SOLUTION OF MULTILAYER... 121

6.3.1 Coupled Strip-Finline Structure

The cross-section of the coupled strip-finline structure is shown in Fig. 6.21. This structure can be viewed as a special case of the structure shown in Fig. 6.20. Since the coupled strip-finline consists of two conductor layers, the spectral domain solution of this structure can be carried out by initially writing Eq. (6.42) in the following form:

$$\begin{bmatrix} G_{11}(\alpha_n,\beta) & G_{12}(\alpha_n,\beta) & G_{13}(\alpha_n,\beta) & G_{14}(\alpha_n,\beta) \\ G_{21}(\alpha_n,\beta) & G_{22}(\alpha_n,\beta) & G_{23}(\alpha_n,\beta) & G_{24}(\alpha_n,\beta) \\ G_{31}(\alpha_n,\beta) & G_{32}(\alpha_n,\beta) & G_{33}(\alpha_n,\beta) & G_{34}(\alpha_n,\beta) \\ G_{41}(\alpha_n,\beta) & G_{42}(\alpha_n,\beta) & G_{43}(\alpha_n,\beta) & G_{44}(\alpha_n,\beta) \end{bmatrix} \begin{bmatrix} \tilde{J}_{x,h} \\ \tilde{J}_{z,h} \\ \tilde{J}_{x,d} \\ \tilde{J}_{z,d} \end{bmatrix} = \begin{bmatrix} \tilde{E}_{x,h} \\ \tilde{E}_{z,h} \\ \tilde{E}_{x,d} \\ \tilde{E}_{z,d} \end{bmatrix} \quad (6.43)$$

However, utilization of the above equation does not lead to an efficient solution. This is because the formation of the final matrix via Eq. (6.43) calls for the expansion of currents associated with fins and the strip. A more efficient solution is possible if Eq. (6.43) is rearranged and written as follows:

$$\begin{bmatrix} I_{11}(\alpha_n,\beta) & I_{12}(\alpha_n,\beta) & I_{13}(\alpha_n,\beta) & I_{14}(\alpha_n,\beta) \\ I_{21}(\alpha_n,\beta) & I_{22}(\alpha_n,\beta) & I_{23}(\alpha_n,\beta) & I_{24}(\alpha_n,\beta) \\ I_{31}(\alpha_n,\beta) & I_{32}(\alpha_n,\beta) & I_{33}(\alpha_n,\beta) & I_{34}(\alpha_n,\beta) \\ I_{41}(\alpha_n,\beta) & I_{42}(\alpha_n,\beta) & I_{43}(\alpha_n,\beta) & I_{44}(\alpha_n,\beta) \end{bmatrix} \begin{bmatrix} \tilde{E}_{z,d} \\ \tilde{E}_{x,d} \\ \tilde{J}_{x,h} \\ \tilde{J}_{z,h} \end{bmatrix} = \begin{bmatrix} \tilde{E}_{z,h} \\ \tilde{E}_{x,h} \\ \tilde{J}_{x,d} \\ \tilde{J}_{z,d} \end{bmatrix} \quad (6.44)$$

It is apparent that in converting the above equation into a final set of homogeneous equations, the current components on the strip $(J_{z,h}$ and $J_{x,h})$ and the field components within the slot $(E_{z,d}$ and $E_{x,d})$ must be expressed in terms of some suitable trial functions. Since in these couplers both the slot and the strip are of small widths, the above arrangement according to section 4.4 should lead to a more economical and efficient solution than that achieved through Eq. (6.43).

So far nothing has been mentioned about elements $I_{11}(\alpha_n,\beta)$, $I_{12}(\alpha_n,\beta)$ etc. of the spectral domain Green's function appearing in Eq. (6.44). These elements can be derived using the technique presented in the previous section. Since this derivation is straightforward but lengthy, for brevity, its details are omitted and only the final results are presented here. Note that $I_{11}(\alpha_n,\beta)$, $I_{12}(\alpha_n,\beta)$ etc. in the following equations are also applicable to Fig. 6.22 which is more general than Fig. 6.21.

$$I_{mn}(\alpha_n,\beta) = \frac{I'_{mn}}{D} \qquad m,n=1,2,3,4 \qquad (6.45.\text{a})$$

$$I'_{11} = -I'_{44} = \beta^2(1 + t_{1,2}^2 t_c)/\cosh(\gamma_{2,n}\delta) \qquad (6.45.\text{b})$$

$$I'_{12} = I'_{21} = -I'_{34} = -I'_{43} = \alpha_n \beta^3 k_{1,2}^2 t_c/\cosh(\gamma_{2,n}\delta) \qquad (6.45.\text{c})$$

$$I'_{13} = I'_{24} = j\alpha_n \beta^3 t(1 + k_2^2 \gamma_{2,n}^2 t_c) \qquad (6.45.\text{d})$$

$$I'_{14} = -j\beta^2 t(k_{2,\beta}^2 + k_2^2 k_{1,\beta}^2 \gamma_{2,n}^2 t_c) \qquad (6.45.\text{e})$$

$$I'_{22} = -I'_{33} = \beta^2(1 + t_{1,2}^2 t_c)/\cosh(\gamma_{2,n}\delta) \qquad (6.45.\text{f})$$

$$I'_{23} = -j\beta^2 t(k_{2,\alpha}^2 + k_2^2 k_{1,\alpha}^2 \gamma_{2,n}^2 t_c) \qquad (6.45.\text{g})$$

$$I'_{31} = I'_{42} = j\beta(E - \alpha_n c' D) \qquad (6.45.\text{h})$$

$$I'_{32} = -j(c'k_{3,\beta}^2 D + B) \qquad (6.45.\text{i})$$

$$I'_{41} = j(\beta^2 F - c' k_{3,\alpha}^2 D) \qquad (6.45.\text{j})$$

where

$$t = \tanh(\gamma_{2n}\delta)/\omega\varepsilon_2 \gamma_{2,n}$$
$$c = \coth(\gamma_{1,n}h)/\omega\mu_1 \gamma_{1,n}$$
$$t_c = tc$$
$$c' = \coth(\gamma_{3,n}d)/\omega\mu_3 \gamma_{3,n}$$

CHAP. 6 SPECTRAL DOMAIN SOLUTION OF MULTILAYER ...

$$E = -\alpha_n \beta^2 [\eta_2 t + c + k_1^2 \gamma_{1,n}^2 C_t + \eta_2 (k_1^2 \gamma_{1,n}^2 + k_{1,2}^4) T_c]$$

$$B = \beta^2 [\eta_2 k_{2,\beta}^2 t + k_{1,\beta}^2 c + k_1^2 k_{2,\beta}^2 \gamma_{1,n}^2 C_t + \eta_2 (\alpha_n^2 \beta^2 k_{1,2}^2 + k_{2,\beta}^2 t_{2,1}^2) T_c]$$

$$F = -[\eta_2 k_{2,\alpha}^2 t + k_{1,\alpha}^2 c + k_1^2 k_{2,\alpha}^2 \gamma_{1,n}^2 C_t + \eta_2 (\alpha_n^2 \beta^2 k_{1,2}^2 + k_{2,\alpha}^2 t_{1,2}^2) T_c]$$

$$D = \beta^2 [1 + (k_2^2 \gamma_{1,n}^2 + k_1^2 \gamma_{2,n}^2) t_c + (k_1 k_2 \gamma_{1,n} \gamma_{2,n} t_c)^2]$$

$$k_i^2 = \omega^2 \mu_i \varepsilon_i \qquad i=1,2,3$$

$$\eta_2 = \varepsilon_2 / \mu_2$$

$$T_c = ct^2$$

$$C_t = c^2 t$$

$$t_{i,j}^2 = \alpha_n^2 k_i^2 + \beta^2 k_j^2 - k_i^2 k_j^2 \qquad i,j=1,2$$

$$k_{i,\beta}^2 = k_i^2 - \beta^2 \qquad i=1,2,3$$

$$k_{i,\alpha}^2 = k_i^2 - \alpha_n^2 \qquad i=1,2,3$$

$$k_{1,2}^2 = k_2^2 - k_1^2$$

$$\gamma_{i,n}^2 = \alpha_n^2 + \beta^2 - k_i^2 \qquad i=1,2,3$$

In Appendix III, there is a computer code in Fortran which can generate all the elements of the spectral domain Green's function introduced in (6.44). Note that for an efficient solution in cases where the slots are much smaller than the strips on the two conductor planes shown in Fig. 6.22, the inverse form of Eq. (6.43) should be employed. The Green's function for these cases can be easily derived using expressions (6.45) given for $I_{11}(\alpha_n, \beta)$, $I_{12}(\alpha_n, \beta)$ etc. The spectral domain analysis of several structures involving fine slots is given in [32]. One such structure is shown in Fig. 6.23.

Let us return to Eq. (6.44) and discuss its solution for the structure shown in Fig. 6.21. As described in the previous section, in order to convert Eq. (6.44) into a set of linear homogeneous equations, we initially require to define some sets of basis functions for expanding $J_{z,h}(x)$, $J_{x,h}(x)$, $E_{z,d}(x)$ and $E_{x,d}(x)$. In the work reported in [29], the trial functions selected are Legendre polynomials and trigonometric functions

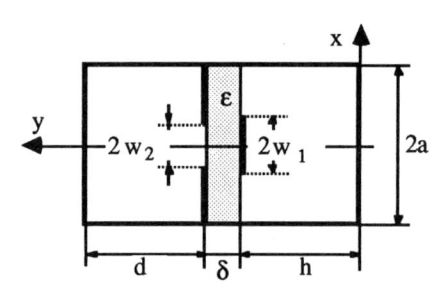

Fig. 6.21 Cross-section of a coupled strip-finline.

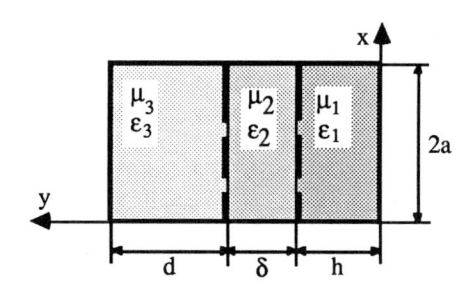

Fig. 6.22 Cross-section of a multiconductor three-layer transmission line.

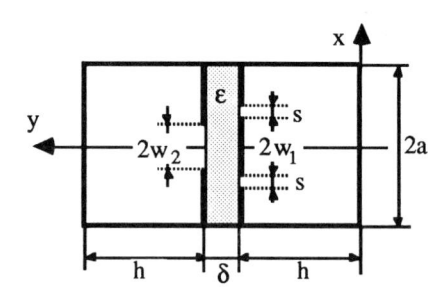

Fig. 6.23 Cross-section of a coupled-slots finline.

and the presented expansions are as follows:

$$J^e_{x,h}(x) \text{ or } E^o_{z,d}(x) = \sum_{p=1}^{P} a_p \sin(\frac{p\pi x}{w}) \qquad |x|<w \qquad (6.46.a)$$

$$J^e_{z,h}(x) \text{ or } E^o_{x,d}(x) = \sum_{q=1}^{Q} b_q P_{2(q-1)}(\frac{x}{w}) \qquad |x|<w \qquad (6.46.b)$$

$$E^e_{z,d}(x) \text{ or } J^o_{x,h}(x) = \sum_{r=1}^{R} c_r \cos(r-\frac{1}{2})\frac{\pi x}{w'} \qquad |x|<w' \qquad (6.46.c)$$

$$E^e_{x,d}(x) \text{ or } J^o_{z,h}(x) = \sum_{s=1}^{S} d_s P_{2s-1}(\frac{x}{w'}) \qquad |x|<w' \qquad (6.46.d)$$

In the above relations, superscripts o and e represent odd and even modes of the coupled strip-finline structure. In the same equations, parameter w and w' are given by:

$$\begin{cases} w = w_2 \\ w' = w_1 \end{cases} \qquad \text{for the odd mode,} \qquad (6.47.a)$$

and by

$$\begin{cases} w = w_1 \\ w' = w_2 \end{cases} \qquad \text{for the even mode.} \qquad (6.47.b)$$

In the transformation of the basis functions, one should be careful to observe that for the even mode (microstrip mode with a magnetic wall plane of symmetry at $x = 0$), the spectrum parameter α_n is:

$$\alpha_n = (n+\frac{1}{2})\frac{\pi}{a} \qquad n=0,1,2,... \qquad (6.48.a)$$

and the same parameter for the odd mode (finline (slot) mode with an electric wall plane of symmetry at $x = 0$) is given by (see Eq. (6.26))

$$\alpha_n = \frac{n\pi}{a} \qquad n=0,1,2,... \qquad (6.48.b)$$

The use of Legendre polynomials and trigonometric functions as suitable basis functions in the spectral domain solution of planar structures is already discussed in section 4.1. These basis functions are also employed by Kawamoto et al [33] in the analysis of coupled microstrip coplanar lines.

CHAP. 6 SPECTRAL DOMAIN SOLUTION OF MULTILAYER ...

Having substituted the transformed forms of Eqs. (6.46) in Eq. (6.44), one can then apply the Galerkin technique to the resulting relation in order to obtain a set of homogeneous equations. By seeking the nontrivial solution of this set, the phase constant can be determined. In connection with the solution of the microstrip line, details of the above procedure for forming the final equations are fully explained in Chapter 3.

For a particular coupled microstrip-finline, phase constants of the even and odd modes computed for different values of w_1 and w_2 are depicted in Fig. 6.24. These results have been obtained using a three-term expansion $(P=Q=R=S=3)$ for each unknown field or current components introduced in Eqs. 6.46. This number of basis functions together with using 150 Fourier terms $(ie: n = 150)$ is sufficient to give an accuracy of better than 0.1% in the computation of β. Note that in the above case, the number of elements in the final matrix is 144. As explained in section 4.2 concerning approximate solutions, a more efficient solution of the coupled microstrip finline is possible at the expense of loss of accuracy. For instance, we may assume that

$$\begin{cases} J_{z,h}(x) = 1 \\ J_{x,h}(x) = 0 \end{cases} \quad -w_1 < x < w_1 \quad (6.49.\text{a})$$

$$\begin{cases} E_{z,d}(x) = 0 \\ E_{x,d}(x) = x \end{cases} \quad -w_2 < x < w_2 \quad (6.49.\text{b})$$

for the microstrip mode and

$$\begin{cases} J_{z,h}(x) = x \\ J_{x,h}(x) = 0 \end{cases} \quad -w_1 < x < w_1 \quad (6.50.\text{a})$$

$$\begin{cases} E_{z,d}(x) = 0 \\ E_{x,d}(x) = 1 \end{cases} \quad -w_2 < x < w_2 \quad (6.50.\text{b})$$

for the slot mode. Using the above approximate fields and currents in the spectral domain formulation of the concerned structure, leads to a characteristic matrix of order

CHAP. 6 SPECTRAL DOMAIN SOLUTION OF MULTILAYER... 127

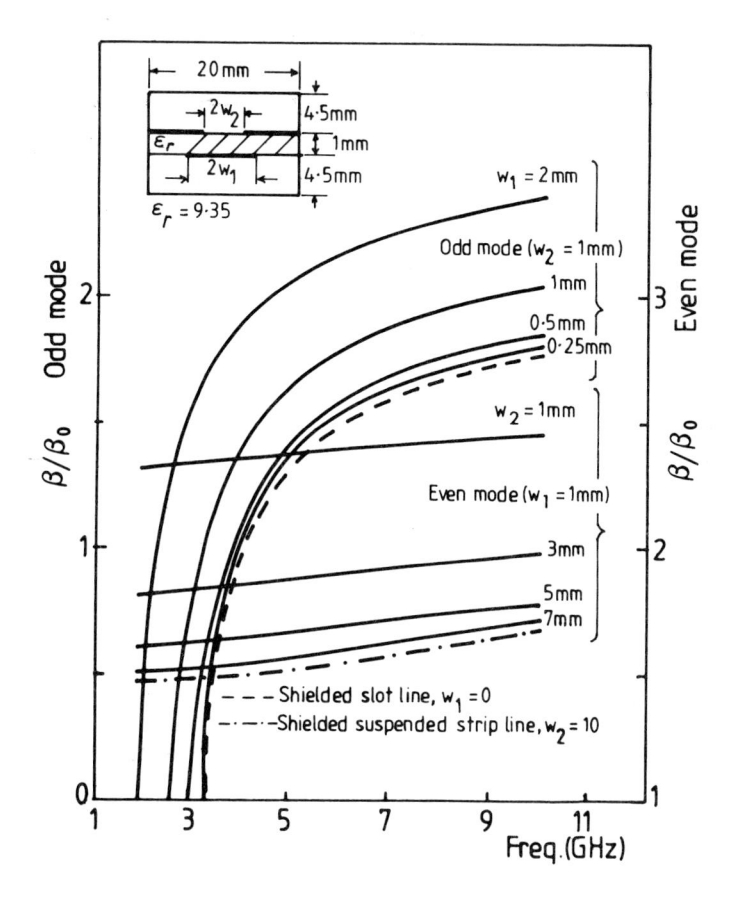

Fig. 6.24 Dispersion characteristics of the even and the odd mode of a shielded microstrip-slot line for different strip and slot sizes computed by the spectral domain technique. (From Mirshekar-Syahkal and Davies [29], Copyright © 1982 IEEE, reproduced with permission.)

2 which is certainly lower than the 12 mentioned earlier. Slightly different approximate forms for fields and currents are given in the work of Ogawa et al [30]. Also in connection with the example studied in this section, the two publications by Sachse et al [34,35] on the suspended microstrip line with tuning septums are recommended for consultation. Associated with this structure, Fig. 6.25 shows the slot fields and the strip currents for the even and odd modes. These results, reported in [35], clearly reflect shapes and relative magnitudes of fields and currents in the slot and on the strip.

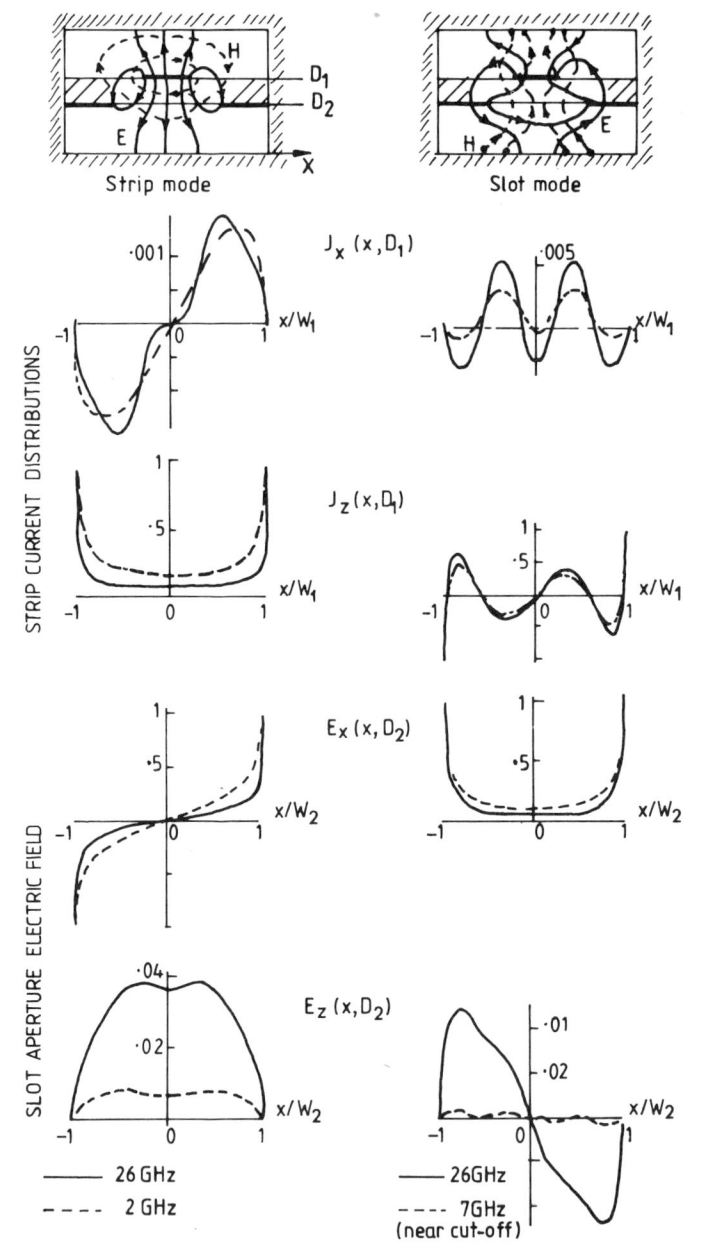

Fig. 6.25 Strip current and slot field distributions corresponding to the even (the strip) mode and to the odd (the slot) mode of a shielded microstrip-slot line (suspended microstrip line with tuning septums) at two different frequencies. (From Sachse et al [35], Copyright © 1980 IEEE, reproduced with permission.)

6.4 Other Approaches for Generating the Spectral Domain Green's Functions of Multilayer Multiconductor Planar Transmission Lines

In the previous sections, the spectral domain formulation of multilayer multiconductor transmission lines was presented using the transfer matrix approach. This approach is very general and applicable to both dynamic (frequency dependent) and static problems alike. However, there are alternative techniques by which the spectral domain Green's function of a multilayer multiconductor planar structure can be generated [10, 36-38]. One of these techniques, known as the spectral domain immittance approach, has received considerable attention recently. This technique, introduced by Itoh [10], basically makes use of the concept of the transverse equivalent transmission line model. The transverse equivalent transmission line model in itself is a developed subject [39] and has been used quite frequently in the past and recent years in order to analyse certain guiding structures [39-41]. But what is interesting in the spectral domain immittance approach is the way that the transverse equivalent transmission line model has been exploited in order to relate the components of currents and electric fields in the Fourier domain. The following example reveals the basic principle behind the spectral domain immittance approach.

Consider the microstrip line shown in Fig. 6.26. In Chapter 3, we discussed a technique to obtain the full-wave spectral domain Green's function of this structure using TE and TM modes. In the spectral domain immittance approach, the same Green's function is derived by employing a different combination of modes known as LSE and LSM modes. For LSE (longitudinal section electric) modes $E_y = 0$ whereas for LSM (longitudinal section magnetic) modes $H_y = 0$ [39]. Analogous to TE and

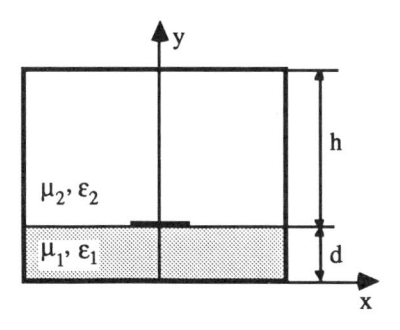

Fig. 6.26 Cross-section of the microstrip line.

CHAP. 6 SPECTRAL DOMAIN SOLUTION OF MULTILAYER...   130

TM modes, LSE and LSM modes can be described in terms of a magnetic potential function $\pi^h(x,y)$ and an electric potential function $\pi^e(x,y)$ respectively. Therefore, for LSE modes,

$$H_y = \pi^h(x,y)e^{-j\beta z} \qquad (6.51.a)$$

and for LSM modes

$$E_y = \pi^e(x,y)e^{-j\beta z} \qquad (6.51.b)$$

With reference to Eqs. (3.3), it is clear that the Fourier expansions of the above expressions in the i th ($i = 1, 2$) dielectric region of the microstrip line are

$$H_{y,i} = \sum_{n=-\infty}^{+\infty} \tilde{\pi}_i^h(\alpha_n, y) e^{-j(\alpha_n x + \beta z)} \qquad (6.52.a)$$

$$E_{y,i} = \sum_{n=-\infty}^{+\infty} \tilde{\pi}_i^e(\alpha_n, y) e^{-j(\alpha_n x + \beta z)} \qquad (6.52.b)$$

From these equations, one can deduce that the field in the microstrip line is the sum of an infinite number of basic LSE and LSM modes. Each mode propagates in the direction v making an angle ξ with the z axis, where

$$\xi = \tan^{-1}(\alpha_n/\beta) \qquad (6.53)$$

Considering an axis u normal to v, it is easy to show that the components of field for a basic (spectral) LSE mode are

$$\tilde{H}_{y,i}, \tilde{E}_{u,i}, \tilde{H}_{v,i} \qquad (6.54)$$

and for a basic (spectral) LSM mode are

$$\tilde{E}_{y,i}, \tilde{H}_{u,i}, \tilde{E}_{v,i} \qquad (6.55)$$

In Fig. 6.27, the direction of the propagation of a spectral LSE/LSM mode as well as the associated field components are shown. Using this figure, where both coordinate systems (x,z) and (v,u) are also depicted, it is easy to deduce the following coordinate transformation:

$$\begin{bmatrix} u \\ v \end{bmatrix} = \begin{bmatrix} -\cos\xi & \sin\xi \\ \sin\xi & \cos\xi \end{bmatrix} \begin{bmatrix} x \\ z \end{bmatrix} \qquad (6.56)$$

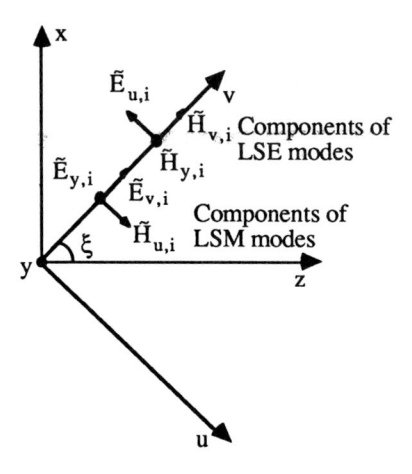

CHAP. 6 SPECTRAL DOMAIN SOLUTION OF MULTILAYER...   131

Fig. 6.27 Coordinates (v,y,u) and (x,y,z) and components of basic LSE and LSM modes.

where:

$$\cos\xi = \frac{\beta}{(\alpha_n^2+\beta^2)^{\frac{1}{2}}} \qquad (6.57.a)$$

$$\sin\xi = \frac{\alpha_n}{(\alpha_n^2+\beta^2)^{\frac{1}{2}}} \qquad (6.57.b)$$

Let us now find out the relations between $\tilde{E}_{u,i}$, $\tilde{H}_{v,i}$ and between $\tilde{H}_{u,i}$, $\tilde{E}_{v,i}$. It is clear that each of these pairs can be imagined as the transverse components of a wave propagating in the y-direction. Therefore, they are related through wave admittances as follows:

$$Y_{i,n}^{LSE} = -\frac{\tilde{H}_{v,i}}{\tilde{E}_{u,i}} = \frac{\gamma_{i,n}}{j\omega\mu_i} \qquad (6.58.a)$$

$$Y_{i,n}^{LSM} = \frac{\tilde{H}_{u,i}}{\tilde{E}_{v,i}} = \frac{j\omega\varepsilon_i}{\gamma_{i,n}} \qquad (6.58.b)$$

where

$$\gamma_{i,n} = (\alpha_n^2+\beta^2-k_i^2)^{\frac{1}{2}} \qquad (6.58.c)$$

CHAP. 6 SPECTRAL DOMAIN SOLUTION OF MULTILAYER ...   132

is the propagation constant in the y direction in the *i* th dielectric region.

Having obtained the above information, we are now in a position to specify the transverse equivalent transmission line models of the microstrip line corresponding to LSE and LSM modes. For each basic LSE and LSM mode, one of the two models shown in Fig. 6.28 is applicable. It is clear that these two models are completely decoupled. This very useful feature is the direct consequence of adopting the *(v,y,u)* coordinate system, Fig. 6.27, for representing the field components of LSE and LSM modes. Note that in the equivalent transmission line models, the short circuits at $y = 0$ and $y = h+d$ represent the ground planes existing at the bottom and top of the microstrip line structures. Also in the same models, $\tilde{J}_u$ and $\tilde{J}_v$ indicate currents on the strip which can be assumed to be generating LSE and LSM modes respectively.

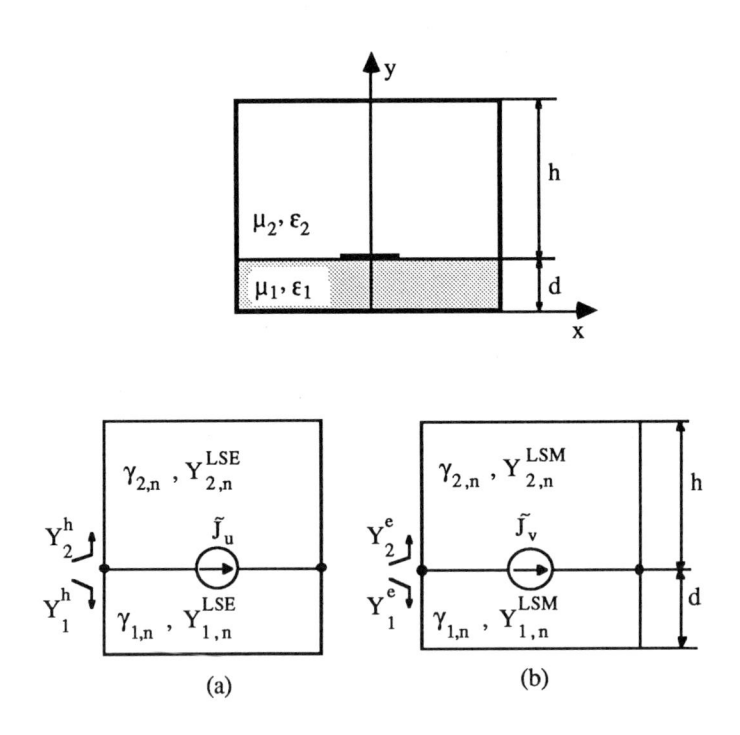

Fig. 6.28 Equivalent transmission line models of the microstrip line; (a) for LSE modes and (b) for LSM modes.

Now by using the transmission line theory, it is possible to express fields $\tilde{E}_u$ and $\tilde{E}_v$ appearing across current sources $\tilde{J}_u$ and $\tilde{J}_v$ in Fig. 6.28 as follows:

$$\tilde{E}_u = Z^h(\alpha_n,\beta)\tilde{J}_u \qquad (6.59.\text{a})$$

$$\tilde{E}_v = Z^e(\alpha_n,\beta)\tilde{J}_v \qquad (6.59.\text{b})$$

where $Z^e(\alpha_n,\beta)$ and $Z^h(\alpha_n,\beta)$ are driving point impedances at $y = d$ for the spectral LSM and LSE modes respectively. These impedances are given by

$$Z^e(\alpha_n,\beta) = \frac{1}{Y^e_{1,n}+Y^e_{2,n}} \qquad (6.60.\text{a})$$

$$Z^h(\alpha_n,\beta) = \frac{1}{Y^h_{1,n}+Y^h_{2,n}} \qquad (6.60.\text{b})$$

where

$$Y^e_{1,n},\ Y^e_{2,n}\ \text{and}\ Y^h_{1,n},\ Y^h_{2,n}$$

are admittances seen at $y = d$ as shown in Fig. 6.28. The expressions for these admittances are readily obtained when the well-known impedance transformation

$$Y = Y_0 \frac{Y_L+Y_0 \tanh(\gamma l)}{Y_0+Y_L \tanh(\gamma l)} \qquad (6.61)$$

is applied to the transmission lines in Fig. 6.28. The results are as follows:

$$Y^e_{2,n} = Y^{LSM}_{2,n} \coth(\gamma_{2,n} h) \qquad (6.62.\text{a})$$

$$Y^e_{1,n} = Y^{LSM}_{1,n} \coth(\gamma_{1,n} d) \qquad (6.62.\text{b})$$

$$Y^h_{2,n} = Y^{LSE}_{2,n} \coth(\gamma_{2,n} h) \qquad (6.62.\text{c})$$

$$Y^h_{1,n} = Y^{LSE}_{1,n} \coth(\gamma_{1,n} d) \qquad (6.62.\text{d})$$

Now let us return to Eqs. (6.59) and rewrite them in the following form [23]:

CHAP. 6 SPECTRAL DOMAIN SOLUTION OF MULTILAYER...

$$\begin{bmatrix} Z^h(\alpha_n,\beta) & 0 \\ 0 & Z^e(\alpha_n,\beta) \end{bmatrix} \begin{bmatrix} \tilde{J}_u \\ \tilde{J}_v \end{bmatrix} = \begin{bmatrix} \tilde{E}_u \\ \tilde{E}_v \end{bmatrix} \quad (6.63)$$

It is not difficult to envisage that the above equation in (v,y,u) coordinate system is equivalent to Eq. (3.45). However, this can be shown explicitly by using the coordinate transformation (6.56). The application of this transformation to the above equation leads to the following expression:

$$\begin{bmatrix} -\cos\xi & \sin\xi \\ \sin\xi & \cos\xi \end{bmatrix} \begin{bmatrix} Z^h(\alpha_n,\beta) & 0 \\ 0 & Z^e(\alpha_n,\beta) \end{bmatrix} \begin{bmatrix} -\cos\xi & \sin\xi \\ \sin\xi & \cos\xi \end{bmatrix} \begin{bmatrix} \tilde{J}_x \\ \tilde{J}_z \end{bmatrix} = \begin{bmatrix} \tilde{E}_x \\ \tilde{E}_z \end{bmatrix} \quad (6.64)$$

or in a more compact form, we obtain

$$\begin{bmatrix} G_{11}(\alpha_n,\beta) & G_{12}(\alpha_n,\beta) \\ G_{21}(\alpha_n,\beta) & G_{22}(\alpha_n,\beta) \end{bmatrix} \begin{bmatrix} \tilde{J}_x \\ \tilde{J}_z \end{bmatrix} = \begin{bmatrix} \tilde{E}_z \\ \tilde{E}_x \end{bmatrix} \quad (6.65)$$

where

$$G_{11}(\alpha_n,\beta) = G_{22}(\alpha_n,\beta) = \sin\xi\cos\xi[Z^e(\alpha_n,\beta) - Z^h(\alpha_n,\beta)] \quad (6.66.a)$$

$$G_{12}(\alpha_n,\beta) = Z^h(\alpha_n,\beta)\sin^2\xi + Z^e(\alpha_n,\beta)\cos^2\xi \quad (6.66.b)$$

$$G_{21}(\alpha_n,\beta) = Z^h(\alpha_n,\beta)\cos^2\xi + Z^e(\alpha_n,\beta)\sin^2\xi \quad (6.66.c)$$

Substituting Eqs. (6.60) and (6.57) in the above equations leads to the determination of elements of the spectral domain Green's function which are already given by Eqs. (3.37).

The spectral domain immittance approach presented above can be readily extended in order to obtain the spectral domain Green's function of a multilayer multiconductor structure. To show this, consider Fig. 6.29, representing the transverse equivalent

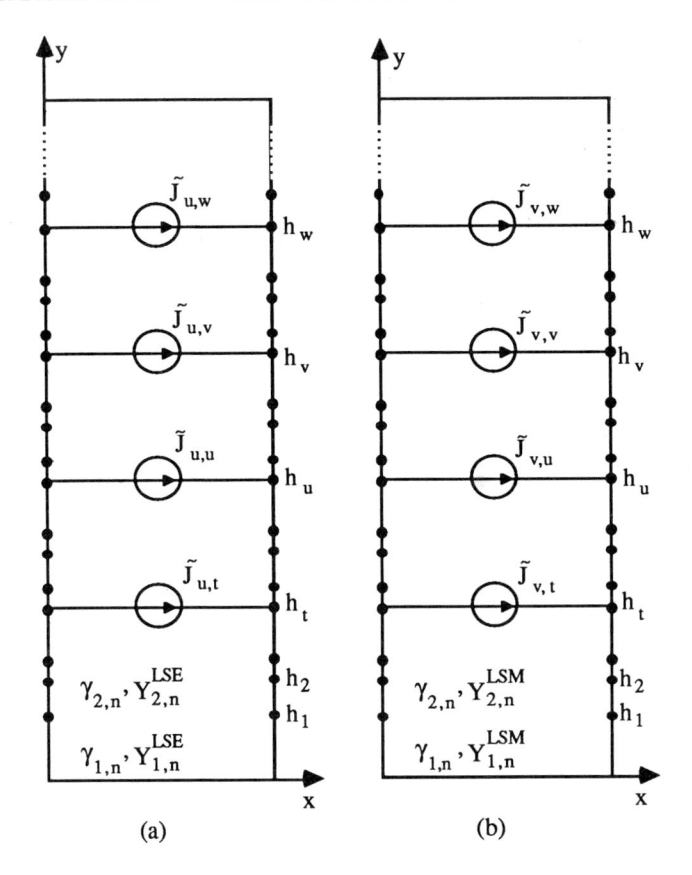

CHAP. 6 SPECTRAL DOMAIN SOLUTION OF MULTILAYER ... 135

Fig. 6.29 Equivalent transmission line models of the multilayer multiconductor planar transmission line shown in Fig. 6.20; (a) for LSE modes and (b) for LSM modes.

transmission line models of Fig. 6.20. From Fig. 6.29.a for the spectral LSE modes, we have

$$\tilde{E}_{u,t} = Z^h_{t,t}(\alpha_n,\beta)\,\tilde{J}_{u,t} + Z^h_{t,u}(\alpha_n,\beta)\tilde{J}_{u,u} + \ldots \qquad (6.67.a)$$

$$\tilde{E}_{u,u} = Z^h_{u,t}(\alpha_n,\beta)\,\tilde{J}_{u,t} + Z^h_{u,u}(\alpha_n,\beta)\tilde{J}_{u,u} + \ldots \qquad (6.67.b)$$

$$\tilde{E}_{u,v} = Z^h_{v,t}(\alpha_n,\beta)\tilde{J}_{u,t} + Z^h_{v,u}(\alpha_n,\beta)\tilde{J}_{u,u} + \ldots \qquad (6.67.c)$$

$$\vdots$$

and from Fig. 6.29.b for the spectral LSM modes, we obtain

$$\tilde{E}_{v,t} = Z^e_{t,t}(\alpha_n,\beta)\tilde{J}_{v,t} + Z^e_{t,u}(\alpha_n,\beta)\tilde{J}_{v,u} + \ldots \quad (6.68.a)$$

$$\tilde{E}_{v,u} = Z^e_{u,t}(\alpha_n,\beta)\tilde{J}_{v,t} + Z^e_{u,u}(\alpha_n,\beta)\tilde{J}_{v,u} + \ldots \quad (6.68.b)$$

$$\tilde{E}_{v,v} = Z^e_{v,t}(\alpha_n,\beta)\tilde{J}_{v,t} + Z^e_{v,u}(\alpha_n,\beta)\tilde{J}_{v,u} + \ldots \quad (6.68.c)$$

$$\vdots$$

where

$$Z^{h/e}_{t,t}(\alpha_n,\beta), \ Z^{h/e}_{u,u}(\alpha_n,\beta), \ \ldots$$

are driving point impedances at $y = h_t$, h_u, ... and terms such as

$$Z^{h/e}_{t,u}(\alpha_n,\beta)$$

represent the transfer impedances expressing the contribution of source at $y = u$ to the field at $y = t$. Note that in the above equations, the second subscripts $t,u,v, \ldots$ in currents and fields denote with which conductor layers these currents and fields are associated and the first subscripts (u and v) in the same parameters denote the u and v axes, Fig. 6.27. Deriving the driving point impedances and transfer impedances is straightforward and in fact they follow using the above definitions. For specific examples, the reader may refer to publications [10,23,42]. Once impedances in Eqs. (6.67) and (6.68) are determined, the desired Green's function can be obtained using the coordinate transformation given by Eq. (6.56).

Recently, based on the equivalent transmission line model, Das et al [36] have also presented a technique for the generation of the spectral domain Green's function of multilayer planar transmission lines. The main feature of the technique is its iterative capability. Furthermore, the method is developed so that it allows the Green's function to be simplified using asymptotic expressions. One other important aspect of the work is the feasibility of finding poles of the Green's functions. These poles are associated with propagation of surface modes in dielectric substrates. In view of the topics presented in this chapter, the work by Das et al [36] can be easily followed. Therefore, we are not going to elaborate it here.

Further works in connection with the derivation of the spectral domain Green's function of planar structures with layers of anisotropic materials are reported in [37,38]. Although, in these works, details of the derivation of the Green's function are somewhat different from those introduced in this chapter, the principles are the same.

References

1. Janiczak B.J., "Multiconductor planar transmission-line structures for high directivity coupler applications", *IEEE*, **MTT-S Digest**, pp.215-218, 1985.

2. Horo M. and Medina F., "Multilayer planar structures for high directivity directional coupler design", *ibid*, pp.283-286, 1986.

3. Paolino D.D., "MIC Overlay coupler design using spectral domain techniques", *IEEE Trans. Microwave Theory Tech.*, **MTT-26**, pp.646-649, 1978.

4. Kitazawa T. and Mittra R., "Quasi-static characteristics of asymmetrical and coupled coplanar-type transmission lines", *ibid*, **MTT-33**, pp.771-778, 1985.

5. Beyer A. and Solbach K., "A new fin-line ferrite isolator for integrated millimeter-wave circuit", *ibid*, **MTT-29**, pp.1344-1348, 1981.

6. Davis L.E. and Sillars D.B., "Millimetric coupled-slot finline components", *ibid*, **MTT-34**, pp.804-808, 1986.

7. Mu T., Ogawa H. and Itoh T., "Characteristics of multiconductor, asymmetric, slow-wave microstrip transmission lines", *ibid*, pp.1471-1477, 1986.

8. Fukuoka Y., Zhang Q., Neikirk D.P. and Itoh T., "Analysis of multilayer interconnection lines for a high-speed digital integrated circuit", *ibid*, **MTT-33**, pp.527-532, 1985.

9. Davies J.B. and Mirshekar-Syahkal D., "Spectral domain solution of arbitrary coplanar transmission lines with multilayer substrate", *ibid*, **MTT-25**, pp.143-146, 1977.

10. Itoh T., "Spectral domain immittance approach for dispersion characteristics of generalised printed transmission lines", *ibid*, **MTT-28**, pp.733-736, 1980.

11. Farrar A. and Adams A.T., "Multilayer microstrip transmission lines", *ibid*, **MTT-22**, pp.889-891, 1974.

12. Horno M. and Medina F., "Accurate approach for computing quasi-static parameters of symmetrical broadside-coupled microstrips in multilayered anisotropic dielectrics", *ibid*, **MTT-34**, pp.729-733, 1986.

13. Wen C.P., "Coplanar waveguide: a surface strip transmission line suitable for nonreciprocal gyromagnetic device applications", *ibid*, **MTT-17**, pp.1087-1090, 1969.

14. Davis M.E., Williams E.W. and Celestini A.C., "Finite boundary correction to the coplanar analysis", *ibid*, **MTT-21**, pp.594-596, 1973.

15. Yamashita E. and Atsuki K.," Analysis of microstrip-like transmission lines by

nonuniform discretization of integral equations", *ibid*, **MTT-24**, pp.195-200, 1976.

16. Knorr J.B. and Kuchler K.D., "Analysis of coupled slots and coplanar strips on dielectric substrate", *ibid*, **MTT-23**, pp.889-891, 1975.

17. Schneider M.V., "Microstrip lines for microwave integrated circuits", *Bell Systems Technical Journal*, **Vol. 48**, No. 5, pp.1421-1444, 1969.

18. Spielman B.E., "Dissipation loss effects in isolated and coupled transmission lines", *IEEE Trans. Microwave Theory Tech.*, **MTT-25**, pp.648-656, 1977.

19. Hopfer S., "The design of ridge waveguides", *IRE Trans. Microwave Theory Tech.*, pp.20-29, 1955.

20. Meier P.J., "Integrated finline millimeter components", *IEEE Trans. Microwave Theory Tech.*, **MTT-22**, pp.1209-1216, 1974.

21. Meier P.J., "Millimeter integrated circuits suspended in the E-plane of rectangular waveguide", *ibid*, **MTT-26**, pp.726-733, 1978.

22. Bates R.N., Nightingale S.J. and Ballard P.M., "Millimeter-wave E-plane components and subsystems", *The Radio and Electronic Engineer*, **Vol. 52**, No. 11/12, pp.506-512, 1982.

23. Baht B. and Koul S., *Analysis, Design and Applications of Finlines*, Artech House, Norwood MA 02062, USA, 1987.

24. Mirshekar-Syahkal D. and Davies J.B., "An accurate, unified solution to various finline structures, of phase constants, characteristic impedance and attenuation", *IEEE Trans. Microwave Theory Tech.*, **MTT-30**, pp.1854-1861, 1982.

25. Pramanick P. and Bhartia P., "Accurate analysis equations and synthesis technique for unilateral finlines", *ibid*, **MTT-33**, pp.24-30, 1985.

26. Knorr J.B. and Shayda P.M., "Millimeter-wave fin-line characteristics", *ibid*, **MTT-28**, pp.737-743, 1980.

27. Schmidt L.P. and Itoh T., "Spectral domain analysis of dominant and higher order modes in finlines", *ibid*, **MTT-28**, pp.981-985, 1980.

28. Mirshekar-Syahkal D. and Jia B., "Analysis of bilateral finline couplers", *Electron. Letts.*, **Vol. 23**, No. 1, pp.577-579, 1987.

29. Mirshekar-Syahkal D. and Davies J.B., "Accurate analysis of coupled strip-finline structure for phase constant, characteristic impedance, dielectric and conductor losses", *IEEE Trans. Microwave Theory Tech.*, **MTT-30**, pp.906-910, 1982.

30. Ogawa H. and Aikawa M., "Analysis of coupled microstrip slot lines", *The Transaction of the IECE of Japan*, **Vol. E62**, No. 4, pp.396-402, 1979.

31. Kawano K., "Hybrid-mode analysis of coupled microstrip-slot resonators", *IEEE*

Trans. Microwave Theory Tech., **MTT-30**, pp.38-41, 1985.

32. Schmidt L.P. and Itoh T., "Characteristics of a generalised fin-line for millimeter-wave integrated circuits". Int. J. of Infrared and Millimeter Waves, **Vol. 2**, No. 3, pp.427-436, 1981.

33. Kawamoto K., Hirota K., Niizaki N., Fujiwara Y. and Ueki K., "Small size VCO module for 900 MHz band using coupled microstrip-coplanar lines", *IEEE, MTT-S Digest*, pp.689-692, 1985.

34. Sachse K. and Citerne J., "Frequency-dependent solution of suspended microstrip line with tuning septums", *Electron. Letts.*, **Vol. 16**, pp.264-266, 1980.

35. Sachse K., Sawicki A., Faucon L., Toutain S., Citerne J., Villotte J.P., Aubroug M. and Garault Y., "Waveguide treatment of the suspended microstrip line with tuning septums using the spectral domain approach and the finite-element method", *IEEE*, **MTT-S**, *Int. Microwave Symp.*, pp.468-470, 1980.

36. Das N.K. and Pozar D.M., "A generalised spectral-domain Green's function for multilayer dielectric substrates with application to multilayer transmission lines", *IEEE Trans. Microwave Theory Tech.*, **MTT-35**, pp.326-335, 1987.

37. Marques R. and Horno M., "On the spectral dyadic Green's function for stratified linear media; application to multilayer MIC lines with anisotropic dielectrics", *IEE Proc.*, **Vol. 134**, Pt. H, No. 3, pp.241-248, 1987.

38. Krowne C.M., "Determination of the Green's function in the spectral domain using a matrix method: Application to radiators or resonators immersed in a complex anisotropic layered media", *IEEE Trans., Antennas and Propagation*, **AP-34**, pp.247-253, 1986.

39. Collin R.E., *Field Theory of Guided Waves*, McGraw Hill, New York, 1960.

40. Saad A.M.K. and Begemann, "Electrical performance of finlines of various configurations", *IEE Journal on Microwaves, Optics and Acoustics,* **Vol. 1**, No. 2, pp.81-88, 1977.

41. Crampagne R., Ahmadpanah M. and Guiraud J., "A simple method for determining the Green's function for a large class of MIC lines having multilayered dielectric structures", *IEEE Trans. Microwave Theory Tech.*, **MTT-26**, pp.82-87, 1978.

42. Chan C.H. and Mittra R., "The propagation characteristics of signal lines embedded in a multilayered structure in the presence of a periodically perforated ground plane", *ibid*, **MTT-36**, pp.968-975, 1988.

Chapter 7

MISCELLANEOUS APPLICATIONS OF THE SPECTRAL DOMAIN TECHNIQUE

In the previous chapters, the spectral domain technique for solving multilayer multiconductor planar transmission lines was gradually developed.

In this chapter, the objective is to demonstrate further capabilities of this technique by reviewing the spectral domain solutions of a selection of planar structures other than transmission lines. Included in this selection are basic examples from planar resonators, radiators, periodic structures, discontinuities, scattering problems etc. As shown in the previous chapter, the effects of extra dielectrics and/or conductors introduced in the basic structure could still be examined using the spectral domain technique. It has to be mentioned that new applications of the spectral domain technique are still emerging. Therefore, it is impossible to present a definitive list of all structures treatable by this powerful technique. However, like the present applications, future applications of the spectral domain technique share the basic principles already elaborated in the earlier chapters in connection with the solution of planar transmission lines.

7.1 Solution of Microstrip-Type Resonators

Depending on the application requirements, various types of planar resonators have been developed and implemented in microwave integrated circuits. Some examples of these resonators, including a microstrip line resonator, a ring resonator, a disk resonator and a triangular patch resonator are shown in Fig. 7.1. The spectral domain analysis of these structures and many similar resonators have already appeared in the literature [1-11]. In this section, we briefly examine the solution of the shielded version of the microstrip line resonator shown in Fig. 7.2. Apart from the fact that the

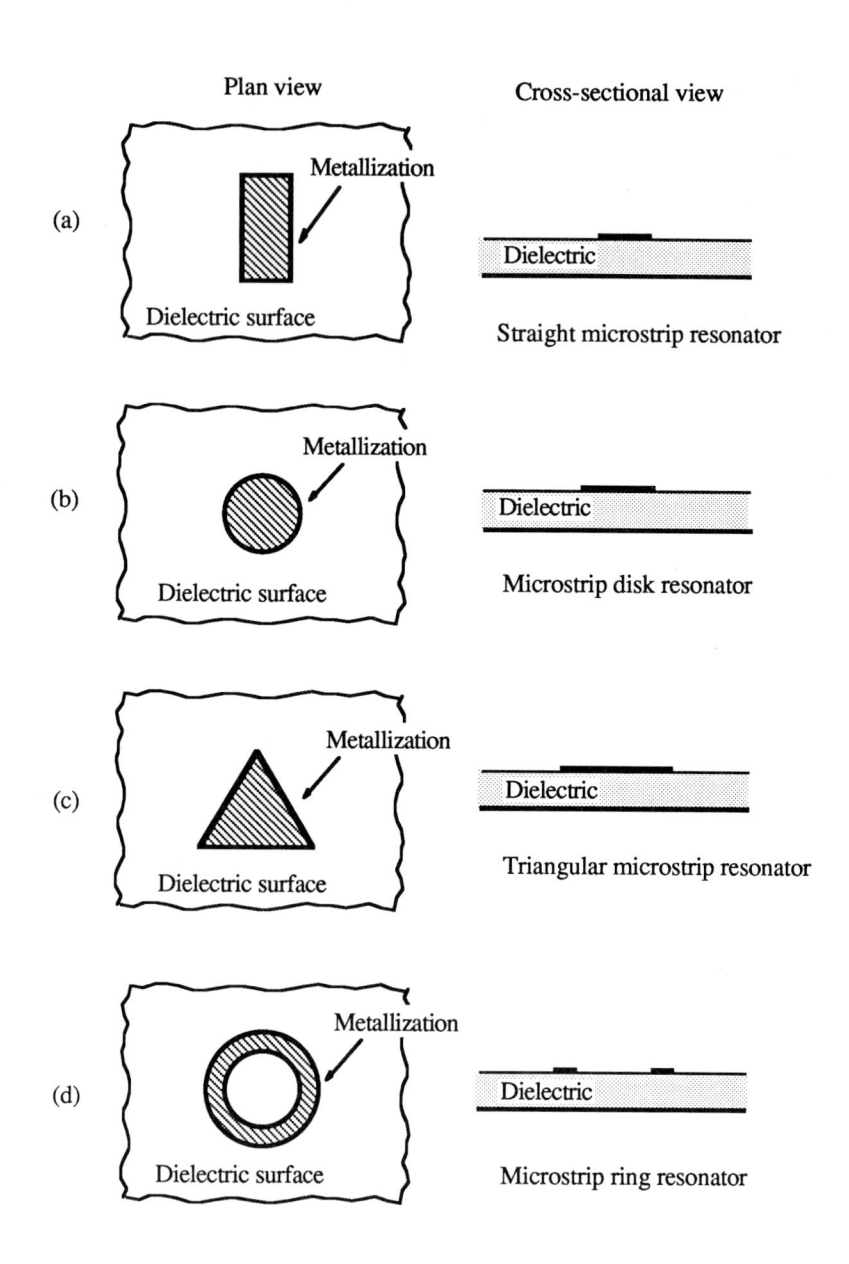

Fig. 7.1 Some examples of planar resonators.

CHAP. 7 MISCELLANEOUS APPLICATIONS OF... 143

microstrip line resonator has the most basic structure compared to similar resonators, its spectral domain formulation is very similar to that presented for the microstrip line. As a comparison, consider a ring or a disk resonator, Fig. 7.1. Because of the circular geometry of a ring or a disk, it is more convenient to make use of the Hankel transform in order to obtain the spectral domain analysis of this type of resonators [10,11]. The Hankel transformation is analogous to the Fourier transform, but it is more appropriate in cylindrical coordinate system.

Now let us return to Fig. 7.2 depicting the shielded microstrip resonator. As will be shown later, the presence of the shield in this structure is essential in order to avoid some of the difficulties associated with calculating resonant frequencies when the

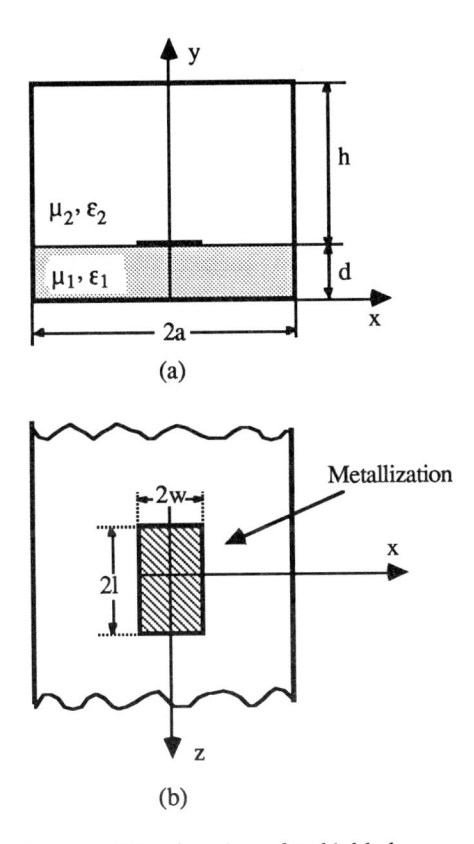

Fig. 7.2 (a) End view and (b) plan view of a shielded microstrip resonator.

resonator is open. The spectral domain solution of the shielded microstrip resonator was originally proposed by Itoh [1]. This technique has later been employed by some investigators to analyse other planar resonators [2-5]. The essence of the technique is as follows.

According to section 2.5, the electromagnetic field in the *i* th dielectric region of the shielded microstrip resonator, Fig. 7.2, can be expressed in terms of two scalar potentials Φ_i and Ψ_i as follows:

$$E_{z,i} = k_i^2 \Phi_i + \frac{\partial^2 \Phi_i}{\partial z^2} \qquad (7.1.a)$$

$$H_{z,i} = k_i^2 \Psi_i + \frac{\partial^2 \Psi_i}{\partial z^2} \qquad (7.1.b)$$

$$E_{x,i} = \frac{\partial^2 \Phi_i}{\partial x \partial z} - j\omega\mu_i \frac{\partial \Psi_i}{\partial y} \qquad (7.1.c)$$

$$H_{x,i} = j\omega\varepsilon_i \frac{\partial \Phi_i}{\partial y} + \frac{\partial^2 \Psi_i}{\partial x \partial z} \qquad (7.1.d)$$

$$E_{y,i} = \frac{\partial^2 \Phi_i}{\partial y \partial z} + j\omega\mu_i \frac{\partial \Psi_i}{\partial x} \qquad (7.1.e)$$

$$H_{y,i} = -j\omega\varepsilon_i \frac{\partial \Phi_i}{\partial x} + \frac{\partial^2 \Psi_i}{\partial y \partial z} \qquad (7.1.f)$$

where $i=1, 2$ and $k_i = \omega^2 \mu_i \varepsilon_i$. Note that the above equations are comparable to Eqs. (2.31) stated for transmission lines. In fact, since the behaviour of Φ_i and Ψ_i with z is explicitly known for transmission lines, ie:

$$\Phi_i = \phi_i(x,y)e^{-j\beta z}, \quad \Psi_i = \psi_i(x,y)e^{-j\beta z},$$

Eqs. (7.1) can be simplified to Eqs. (2.31).

As mentioned in section 2.5, Φ_i and Ψ_i in Eqs. (7.1) are solutions of the wave equation

$$\nabla^2 u_i + k_i^2 u_i = 0 \qquad (7.2)$$

subject to boundary conditions. In its expanded form, Eq. (7.2) can be written as follows:

$$\frac{\partial^2 u_i}{\partial x^2} + \frac{\partial^2 u_i}{\partial y^2} + \frac{\partial^2 u_i}{\partial z^2} + k_i^2 u_i = 0 \tag{7.3}$$

Now by using the concept of the Fourier transform introduced in section 3.2.1, it is possible to obtain the equivalent form of Eq. (7.3) in the Fourier domain. This is given by

$$\frac{d^2 \tilde{u}_i}{dy^2} - \gamma_{i,n}^2 \tilde{u}_i = 0 \tag{7.4}$$

where

$$\gamma_{i,n}^2 = \alpha_n^2 + \beta^2 - k_i^2 \tag{7.5}$$

and

$$\tilde{u}_i = \frac{1}{4\pi a} \int_{-\infty}^{+\infty} \int_{-a}^{+a} u_i e^{j(\alpha_n x + \beta z)} dx dz \tag{7.6}$$

In view of Fig. 7.2, the application of the two dimensional Fourier transform (7.6) is clearly justified. Due to the existence of side walls in the structure of concern, the spectral parameter α_n in the above equation has discrete values given by

$$\alpha_n = (n+\tfrac{1}{2})\frac{\pi}{a} \qquad n = 0,1,2,\ldots \tag{7.7}$$

for the even mode excitation of the resonator. Parameter β, on the other hand, is continuous, varying from $-\infty$ to $+\infty$.

Since Eq. (7.4) is identical to Eq. (3.12), its general solution is given by Eq. (3.15). Thus, in the Fourier domain, Φ_i and Ψ_i are given by Eqs. (3.16) and (3.17). If we substitute these potential functions in the transformed forms of Eqs. (7.1) and follow the same procedure established in the spectral domain formulation of the microstrip line, we arrive at the spectral domain Green's function of the resonator:

$$\begin{bmatrix} G_{11}(\alpha_n,\beta,\omega) & G_{12}(\alpha_n,\beta,\omega) \\ G_{21}(\alpha_n,\beta,\omega) & G_{22}(\alpha_n,\beta,\omega) \end{bmatrix} \begin{bmatrix} \tilde{J}_x \\ \tilde{J}_z \end{bmatrix} = \begin{bmatrix} \tilde{E}_z \\ \tilde{E}_x \end{bmatrix} \tag{7.8}$$

It is interesting to note that the spectral domain Green's function of the microstrip resonator is the same as that of the microstrip line, given by Eqs. (3.37). This useful feature, in fact, is a direct result of the application of the Fourier transform in the z-direction, decomposing a complicated wave into infinite number of basic waves propagating in the z-direction.

As explained in section 3.3.2.2, Eq. (7.8) can be converted to a system of linear homogeneous equations. To do this, the current distributions in Eq. (7.8) are initially written in terms of series of basis functions. The Galerkin technique is then applied. Although the resulting system of equations is the same as that given by Eqs. (3.66), its coefficients are formed differently. For instance, compare Eq. (3.67.a) with the following equation representing one of the coefficients

$$C_{q',p}^{1,1}(\omega) = \sum_{n=-\infty}^{+\infty} \int_{-\infty}^{+\infty} G_{11}(\alpha_n,\beta,\omega) \tilde{J}_{z,q'}^*(\alpha_n,\beta) \tilde{J}_{x,p}(\alpha_n,\beta) d\beta \qquad (7.9)$$

From the resulting system of equations, it is readily concluded that the resonant frequency ω which is unknown, can be determined by setting the determinant of the coefficients equal to zero (see section 3.3.2.1).

As in the case of transmission lines, we have a broad choice of basis functions. However, since current distributions for resonant-type structures have two-dimensional variations, the selection of the basis functions in particular should be made with extreme care. Otherwise, the solution in respect of computer time becomes unwieldy. This can be easily recognised from Eq. (7.9) which represents a typical coefficient. In fact, due to the double summation involved in Eq. (7.9), its evaluation can be very time consuming.

In order to obtain the resonant frequency of the fundamental mode of the microstrip resonator, Itoh [1] assumes that the current distributions can be represented as

$$J_x(x,z) = a_1 j_x(x) i_x(z) \qquad (7.10.a)$$

$$J_z(x,z) = b_1 j_z(x) i_z(z) \qquad (7.10.b)$$

where for $j_x(x)$, $i_x(z)$, $j_z(x)$ and $i_z(z)$ the following expressions which closely approximate the actual current distributions, are considered:

$$j_x(x) = \frac{1}{w}\sin\left(\frac{\pi x}{w}\right) \qquad |x| < w \qquad (7.11.a)$$

$$i_x(z) = \frac{z}{2l^2} \qquad |z| < l \qquad (7.11.b)$$

$$j_z(x) = \frac{1}{2w}\left(1+\left|\frac{x}{w}\right|^3\right) \qquad |x| < w \qquad (7.11.c)$$

$$i_z(z) = \frac{1}{l}\cos\left(\frac{\pi z}{2l}\right) \qquad |z| < l \qquad (7.11.d)$$

The Fourier transforms of the above functions are as follows:

$$\tilde{j}_x(\alpha_n) = \frac{2\pi \sin(\alpha_n w)}{(\alpha_n w)^2 - \pi^2} \qquad (7.12.a)$$

$$\tilde{i}_x(\beta) = \frac{\cos(\beta l)}{\beta l} - \frac{\sin(\beta l)}{(\beta l)^2} \qquad (7.12.b)$$

$$\tilde{j}_z(\alpha_n) = \frac{2\sin(\alpha_n w)}{\alpha_n w} + \frac{3}{(\alpha_n w)^2}\left\{\cos(\alpha_n w) - \frac{2\sin(\alpha_n w)}{\alpha_n w} + \frac{2[1-\cos(\alpha_n w)]}{(\alpha_n w)^2}\right\} \qquad (7.12.c)$$

$$\tilde{i}_z(\beta) = -\frac{\pi \cos(\beta l)}{(\beta l)^2 - \left(\frac{\pi}{2}\right)^2} \qquad (7.12.d)$$

Thus, the Fourier transforms of Eqs. (7.10.a) and (7.10.b) are

$$\tilde{J}_x = a_1 \tilde{j}_x(\alpha_n) \tilde{i}_x(\beta) \qquad (7.13.a)$$

$$\tilde{J}_z = b_1 \tilde{j}_z(\alpha_n) \tilde{i}_z(\beta) \qquad (7.13.b)$$

It is clear that by substituting Eqs. (7.13) in Eq. (7.8) and applying the Galerkin technique to the resulting equation, one obtains a system of equations similar to Eqs. (4.11). Therefore, the characteristic equation for the resonator resembles Eq. (4.13) from which the resonant frequency can be computed.

At the beginning of this section, it has been briefly mentioned that the presence of the shield in Fig. 7.2 relaxes mathematical complexities involved in the evaluation of resonant frequencies of open planar resonators. To be more explicit, it should be said that the computation of the elements of the characteristic matrix becomes simpler if the width of the shield is so that no wave is propagating in the z-direction for all values of

β. To elaborate on this fact, consider equation

$$u_i = \sum_n \int_{-\infty}^{+\infty} \tilde{u}_i e^{-j(\alpha_n x + \beta z)} d\beta \tag{7.14}$$

representing the inverse Fourier transform for the Fourier transform defined by Eq. (7.6). The physical interpretation of Eq. (7.14) is clear; the field inside the resonator consists of fields due to plane waves. The Fourier parameter β particularly denotes phase constants of waves travelling in the z-direction in the structure. Depending on whether a wave is propagating or evanescent, its corresponding β will be real or imaginary. Generally speaking, β can be considered to be a complex variable. This becomes more meaningful when the integral transform in Eq. (7.14) is interpreted as the inverse of a double-sided Laplace transform, in which case β is a complex variable. In view of the above discussion, let us examine one of the coefficients of the characteristic matrix. For instance, consider Eq. (7.9). For some values of β, $G_{11}(\alpha_n,\beta,\omega)$ in this equation becomes infinite. These poles are found by setting the denominator of $G_{11}(\alpha_n,\beta,\omega)$, which is Eq. (3.37.d), equal to zero. Note that Eq. (3.37.d) appears in all the coefficients and it represents the eigenvalue equation for modes in the dielectric loaded waveguide shown in Fig. 7.3. When this waveguide is at cut-off for all its modes, all the poles are imaginary and hence all modes are evanescent. As a result, without requiring a special treatment, the integral in Eq. (7.9) can be taken along the real β axis and its convergence is guaranteed. However, if the size of the shield is such that some of the poles are real, the path of the integration along the real β axis should be indented and should circle around the singular points. This obviously adds to the complexity in the numerical evaluation of the coefficients.

In the spectral domain analysis of open planar resonators, a situation similar to the

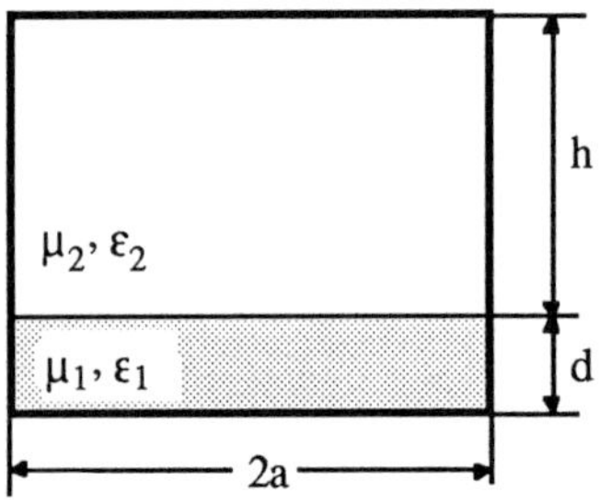

Fig. 7.3 Cross-section of the dielectric loaded waveguide.

CHAP. 7 MISCELLANEOUS APPLICATIONS OF ... 149

above is encountered. Real poles in open resonators are associated with the excitation of surface waves in the substrate. It should be further noted that in the analysis of open resonators, double integrals in terms of α and β appear in the expressions for the coefficients in the final linear system of equations. For example, the equivalent expression to Eq. (7.9) for the open resonator shown in Fig. 7.1.a is

$$C_{q',p}^{l,l}(\omega) = \int_{-\infty}^{+\infty}\int_{-\infty}^{+\infty} G_{11}(\alpha,\beta,\omega)\tilde{J}_{z,q'}^{*}(\alpha,\beta)\tilde{J}_{x,p}(\alpha,\beta)d\alpha d\beta \qquad (7.15)$$

where $G_{11}(\alpha,\beta,\omega)$ is expressed in a suitable form for Fig. 7.1.a. Details of the evaluation of the above integral as well as those of the surface waves whose effects appear in Eq. (7.15) can be found in many references including [12-17].

The technique proposed by Itoh [1] which was explained earlier, has been the basis for the analysis of many planar resonators [2-5]. One of these planar resonators whose analysis is particularly instructive from the current distribution representation point of view, is the isosceles triangular patch resonator, Fig. 7.4. For the fundamental mode of this resonator, the components of the current distribution $J_x(x,z)$ and $J_z(x,z)$ can be approximated as follows [5]:

$$J_x(x,z) \approx 0 \qquad (7.16.a)$$

$$J_z(x,z) \approx (1+|\frac{x}{w_o}|^3)\sin\left(\frac{\pi z}{l}\right) \qquad \text{over the patch} \qquad (7.16.b)$$

where w_o and l are shown in Fig. 7.4. The Fourier transform of $J_z(x,z)$ is found through the following expression:

$$\tilde{J}_z = \frac{1}{2\pi a}\int_0^l \sin\left(\frac{\pi z}{l}\right)e^{j\beta z}\int_0^{w(z)}(1+|\frac{x}{w_o}|^3)\cos(\alpha_n x)dxdz \qquad (7.17.a)$$

where

$$w(z) = \frac{w_o}{l}(l-z) \qquad (7.17.b)$$

It should be noted that although in the case of the triangular patch resonator, omission of $J_x(x,z)$ from the analysis is expected to lead to a gross error in the evaluation of ω, in practice it is surprising to find how closely the experimental and the theoretical resonant frequencies are correlated [5].

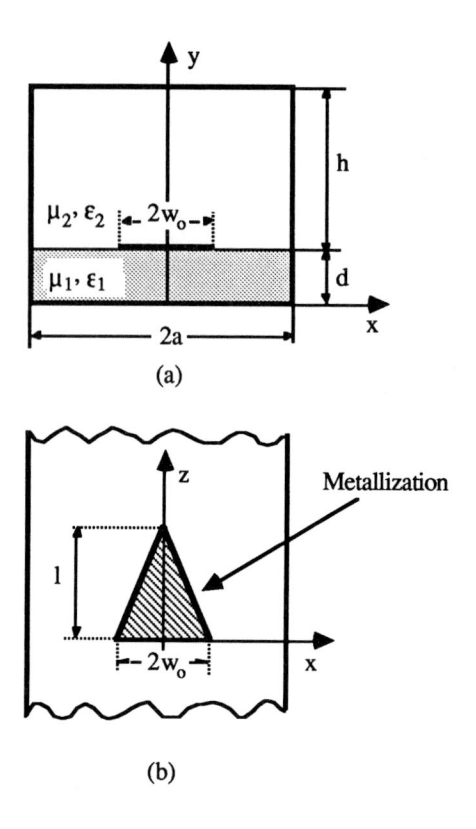

Fig. 7.4 (a) End view and (b) plan view of an isosceles triangular patch resonator.

7.2 Solution of Microstrip Patch Antennas

From the structural configuration point of view, microstrip patch antennas resemble microstrip patch resonators. The main difference between these two planar components, however, lies in the way they are designed to perform their functions properly; namely, a microstrip antenna is designed for maximum radiation whereas in the design of a patch resonator, radiation has to be minimized. In Fig. 7.5, some microstrip-type antennas as well as several other planar radiators are shown.

Due to certain advantages of planar antennas, the past decade has witnessed a rapid

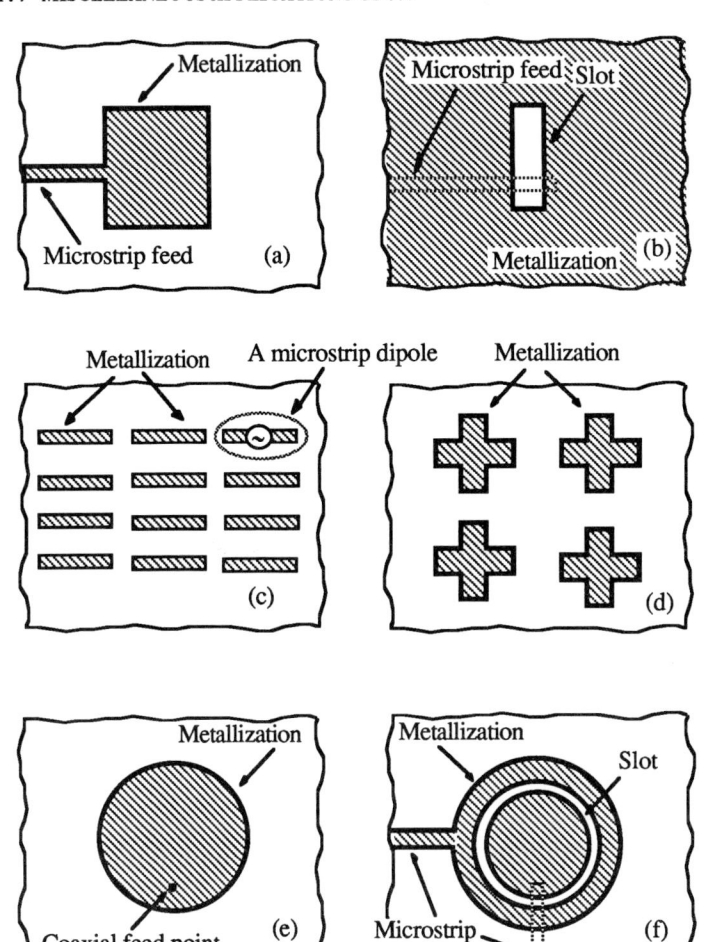

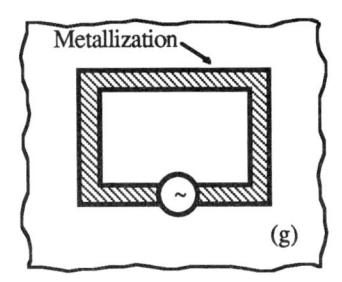

Fig. 7.5 Plan views of some planar radiators made on dielectric substrates; (a) a microstrip patch antenna, (b) a slot antenna fed by the microstrip line, (c) an array of microstrip dipoles, (d) an array of microstrip cross-dipoles, (e) a circular microstrip antenna fed by a coaxial line, (f) a dual-frequency microstrip-slot ring antenna and (g) a rectangular loop microstrip antenna.

use of these radiators in communication and radar systems [18,19]. Along with this development, a rapid progress in mathematical modelling and accurate design of planar antennas has been reported in the literature. This section briefly discusses one particular modelling technique which is based on the spectral domain method. Although this technique is applicable to a wide variety of planar radiators, we only examine the spectral domain solution of the rectangular microstrip antenna below. For further examples of the applications of this technique, the reader may refer to references [20-24].

Fig. 7.6 shows a rectangular microstrip antenna excited by a coaxial feed probe. The spectral domain analysis of this structure can be carried out with or without considering the probe [25,26]. The choice depends on whether both the resonant frequency and the input impedance of the antenna are required to be assessed, or whether our major concern is the evaluation of the resonant frequency of the antenna, only. We briefly examine the solution in each case.

When no probe is considered in the structure, the geometry of the rectangular patch antenna becomes identical to that of the rectangular patch resonator shown in Fig. 7.1.a. Hence, the resonant frequency of the antenna can be obtained using the spectral domain formulation discussed in section 7.1 for planar resonators. Note that for an open rectangular planar resonator, the Fourier transform is two-dimensional and can be defined by the following equations:

$$\tilde{u}_i = \frac{1}{(2\pi)^2} \int_{-\infty}^{+\infty}\int_{-\infty}^{+\infty} u_i e^{j(\alpha x + \beta z)} dx\, dz \qquad (7.18.\text{a})$$

$$u_i = \int_{-\infty}^{+\infty}\int_{-\infty}^{+\infty} \tilde{u}_i e^{-j(\alpha x + \beta z)} d\alpha\, d\beta \qquad (7.18.\text{b})$$

In the above equations, x and z are real variables whereas α and β are real or imaginary. These latter parameters can be generally assumed to be complex variables in which case Eqs. (7.18) can represent the two-sided Laplace transform. Also note that for open resonators, Eq. (7.8) reads as follows:

$$\begin{bmatrix} G_{11}(\alpha,\beta,\omega) & G_{12}(\alpha,\beta,\omega) \\ G_{21}(\alpha,\beta,\omega) & G_{22}(\alpha,\beta,\omega) \end{bmatrix} \begin{bmatrix} \tilde{J}_x \\ \tilde{J}_z \end{bmatrix} = \begin{bmatrix} \tilde{E}_z \\ \tilde{E}_x \end{bmatrix} \qquad (7.19)$$

CHAP. 7 MISCELLANEOUS APPLICATIONS OF ...   153

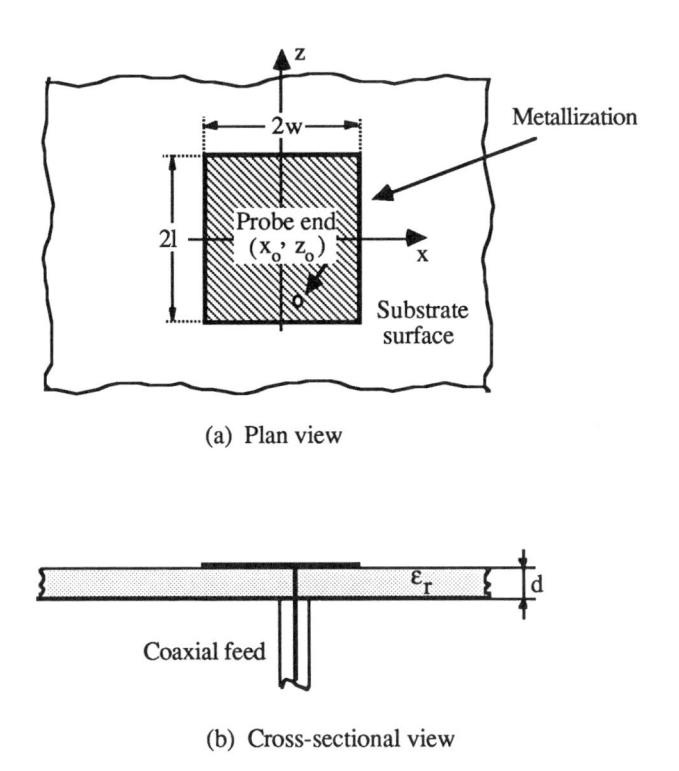

Fig. 7.6 A rectangular patch antenna with coaxial probe feed.

where the matrix elements $G_{11}(\alpha,\beta,\omega)$, $G_{12}(\alpha,\beta,\omega)$ etc. can be generated from those in Eq. (3.37) by letting $h \to \infty$ and replacing α_n by α.

Analogous to the example treated in the previous section, in order to obtain the resonant frequency of a source-free patch resonator, Eq. (7.19) is converted to a set of homogeneous linear equations. The unknowns in the resulting system of equations are the amplitudes of the basis currents, and the coefficients involved in the system are given by some integral expressions. A typical coefficient is given in Eq. (7.15). Since the system of equations is homogeneous, its nontrivial solution is achieved when the determinant of the coefficients vanishes. This occurs at the resonant frequencies of the resonator.

Examining Eq. (7.15) which represents a typical coefficient, leads to the conclusion that this coefficient should be generally complex. This is because, in this equation, parameters α and β span the whole spectrum of modes (evanescent and propagating). As a result, resonant frequencies computed should be complex too, ie:

$$\omega = \omega_r + j\omega_i$$

where the real part is the actual resonant frequency of the patch resonator.

The concept of complex resonant frequency, which was initially introduced by Itoh et al [25] in the calculation of resonant frequencies and Q-factors of planar patch antennas, has been the basis for the analysis of several microstrip-type radiators; for instance see [23,27]. Unfortunately in all those works, it appears that the integrals representing the coefficients of the final equations have been evaluated along the real axis (a typical coefficient is given in Eq. (7.15)). Although under certain conditions, ie: thin substrate, low frequency and low permittivity [16], this type of integration is acceptable, generally, the procedure adopted in [23,25,27] for obtaining the values of the coefficients is not accurate. The problem associated with the procedure is addressed in a recent publication by Assailly et al [15] where they present the correct path of integration in order to avoid surface wave poles on the real axis.

Now let us consider the case where the excitation probe is part of the rectangular microstrip antenna, Fig. 7.6. For an electrically thin substrate, the probe may be assumed to be a linear current source [28] producing an incident field onto the antenna metal patch. This field induces a current on the patch whose field together with the incident field make up the total field of the antenna. At the air-dielectric interface, the spectral domain relation between the patch current and the tangential electric field due to it, is already known and is given by Eq. (7.19). Therefore, if $E_{x,I}$ and $E_{z,I}$, representing the tangential components of the incident electric field at the interface, are added to Eq. (7.19), the resulting expressions are the Fourier transforms of $E_{x,T}$ and $E_{z,T}$ of the total field at the interface, ie:

$$\begin{bmatrix} G_{11}(\alpha,\beta) & G_{12}(\alpha,\beta) \\ G_{21}(\alpha,\beta) & G_{22}(\alpha,\beta) \end{bmatrix} \begin{bmatrix} \tilde{J}_x \\ \tilde{J}_z \end{bmatrix} + \begin{bmatrix} \tilde{E}_{z,I} \\ \tilde{E}_{x,I} \end{bmatrix} = \begin{bmatrix} \tilde{E}_{z,T} \\ \tilde{E}_{x,T} \end{bmatrix} \qquad (7.20)$$

Now let us examine how the above equation can be converted to a set of linear equations. As frequently explained in this book, the spectral domain analysis of a planar structure depends on the successful formation of these equations. For the

CHAP. 7 MISCELLANEOUS APPLICATIONS OF ...

brevity of the following discussion, we assume that J_x is vanishingly small. This approximation is allowed when the frequency is low and the antenna is operating at its fundamental mode.

In order to set up a set of linear equations from Eq. (7.20), we follow the same procedure adopted in section 3.3.2.2. This requires initially the expansion of J_z in terms of a set of basis functions (note that $J_x = 0$)

$$J_z = \sum_q b_q J_{z,q} \qquad (7.21)$$

where $J_{z,q}$ is a function of x and z. The linear system, (7.22), can then be set up by substituting the Fourier transform of Eq. (7.21) in Eq. (7.20) and applying the Galerkin technique to the resulting equation:

$$\sum_q b_q < G_{12}(\alpha,\beta) \tilde{J}_{z,q}, \tilde{J}_{z,q'} > = - < \tilde{E}_{z,I}, \tilde{J}_{z,q'} > \quad q'=1,2,... \qquad (7.22)$$

In the above system of equations, the inner products on the left-hand side represent coefficients of the linear equations and are defined by the following integral expression:

$$< G_{12}(\alpha,\beta) \tilde{J}_{z,q}, \tilde{J}_{z,q'} > = \int_{-\infty}^{+\infty} \int_{-\infty}^{+\infty} G_{12}(\alpha,\beta) \tilde{J}_{z,q} \tilde{J}_{z,q'}^* \, d\alpha d\beta \qquad (7.23)$$

In order to evaluate the inner product on the right-hand side of Eq. (7.22), an expression for $E_{z,I}$ should be established first. However, a simple alternative technique is to use the reciprocity theorem [29] which replaces this inner product by an equivalent expression which is easier to compute. To this end, from Parseval's identity [Appendix I], we observe that

$$< \tilde{E}_{z,I}, \tilde{J}_{z,q'} > = \int_{-\infty}^{+\infty} \int_{-\infty}^{+\infty} \tilde{E}_{z,I} \tilde{J}_{z,q'}^* \, d\alpha d\beta = \frac{1}{(2\pi)^2} \int_{-\infty}^{+\infty} \int_{-\infty}^{+\infty} E_{z,I} J_{z,q'} \, dxdz \qquad (7.24)$$

Applying the reciprocity principle to the right-hand side of the above equation leads to the following equation:

CHAP. 7 MISCELLANEOUS APPLICATIONS OF . . .   156

$$\int_{-\infty}^{+\infty}\int_{-\infty}^{+\infty} E_{z,I} J_{z,q'} \, dxdz = \int_{0}^{d}\int_{-\infty}^{+\infty}\int_{-\infty}^{+\infty} i_y E_{y,I(q')} \, dxdy\,dz \quad (7.25)$$

where i_y is the current distribution on the excitation probe and $E_{y,I(q')}$ is the field along the probe due to the basis current $J_{z,q'}$. From the above equation, it is not difficult to realise that the reciprocity theorem allows us to replace the integral over the scatterer to an integral over the source. As assumed earlier, the probe is a linear current source. Thus, i_y is given as

$$i_y = I\delta(x-x_o)\delta(z-z_o) \quad (7.26)$$

where x_o and z_o are the coordinates of the probe; see Fig. 7.6. In the above equation, we can assume that $I = 1\,A$, since I is an arbitrary constant which represents the magnitude of the current flowing in the probe. If we now substitute Eq. (7.26) in Eq. (7.25), we obtain the following equation:

$$<\tilde{E}_{z,I}, \tilde{J}_{z,q'}> = \frac{1}{(2\pi)^2} \int_0^d E_{y,I(q')}(x_o, y, z_o) \, dy \quad (7.27)$$

Considering the fact that $E_{y,I(q')}(x_o,y,z_o)$ is the field in the substrate, induced along the probe due to basis current $J_{q'}$ on the patch, its expression can be found by back substitution. This is explained as follows. Use Eq. (7.19) to find the spectral domain expressions of E_x and E_z in terms of the Fourier transform of J_z. (Note that we have assumed $J_x = 0$). The result is:

$$\tilde{E}_z = G_{12}(\alpha,\beta)\tilde{J}_z \quad (7.28.a)$$

$$\tilde{E}_x = G_{22}(\alpha,\beta)\tilde{J}_z \quad (7.28.b)$$

Then use these equations in conjunction with the Fourier transforms of Eqs. (7.1.a) and (7.1.c) to obtain the unknown potential function coefficients. With reference to Eqs. (3.24), it can be easily deduced that in the Fourier domain, Eqs. (7.1.a) and (7.1.c) in the substrate are given by:

$$\tilde{E}_{z,1} = (k_1^2 - \beta^2) A_1(\alpha,\beta) \sinh(\gamma_1 y) \quad (7.29.a)$$

$$\tilde{E}_{x,1} = -[\alpha\beta A_1(\alpha,\beta) + j\omega\mu_1\gamma_1 C_1(\alpha,\beta)] \sinh(\gamma_1 y) \quad (7.29.b)$$

CHAP. 7 MISCELLANEOUS APPLICATIONS OF ...   157

Bearing in mind that at $y = d$, Eq. (7.28.a) is equal to Eq. (7.29.a), and Eq. (7.28.b) is equal to Eq. (7.29.b), it is easy to show that the potential function coefficients are given by expressions:

$$A_l(\alpha,\beta) = \frac{G_{12}(\alpha,\beta)\tilde{J}_z}{(k_1^2-\beta^2)\sinh(\gamma_1 d)} \qquad (7.30.a)$$

$$C_l(\alpha,\beta) = j\frac{\tilde{J}_z}{\omega\mu_1\gamma_1\sinh(\gamma_1 d)}\left[G_{22}(\alpha,\beta) + \frac{\alpha\beta}{(k_1^2-\beta^2)}G_{12}(\alpha,\beta)\right] \qquad (7.30.b)$$

Now, substitute the above coefficients in the Fourier transform of $E_{y,l}$, given by equation

$$\tilde{E}_{y,l} = -j[\gamma_1\beta A_l(\alpha,\beta) + j\omega\mu_1\alpha C_l(\alpha,\beta)]\cosh(\gamma_1 y) \qquad (7.31)$$

to obtain the following expression

$$\tilde{E}_{y,l} = j\left[\frac{\beta}{\gamma_1}G_{12}(\alpha,\beta) + \frac{\alpha}{\gamma_1}G_{22}(\alpha,\beta)\right]\tilde{J}_z\cosh(\gamma_1 y)/\sinh(\gamma_1 d) \qquad (7.32)$$

Note that Eq. (7.31) is obtained in a similar way that Eq. (3.24.b) has been derived.

From Eq. (7.32), it is now easy to find $E_{y,l(q')}$ due to $J_{z,q'}$. This is given by the following equation:

$$E_{y,l(q')} = \int_{-\infty}^{+\infty}\int_{-\infty}^{+\infty}\tilde{e}_{y,q'}\frac{\cosh(\gamma_1 y)}{\cosh(\gamma_1 d)}e^{-j(\alpha x+\beta z)}d\alpha d\beta \qquad (7.33)$$

where

$$\tilde{e}_{y,q'} = j\left[\frac{\beta}{\gamma_1}G_{12}(\alpha,\beta) + \frac{\alpha}{\gamma_1}G_{22}(\alpha,\beta)\right]\tilde{J}_{z,q'}\coth(\gamma_1 d) \qquad (7.34)$$

Therefore, by substituting Eq. (7.33) in Eq. (7.27), the evaluation of the desired inner products on the right-hand side of Eq. (7.22) becomes possible. As a result, the expansion coefficients $b_1, b_2, ...$ can be computed from Eq. (7.22) simply by a matrix inversion. Having determined these coefficients, the input impedance of the antenna, Z_{in}, can be computed using the expression:

$$Z_{in} = \frac{V}{I} = -\int_0^d E_{y,l}(x_o,y,z_o)dy \qquad (7.35)$$

CHAP. 7 MISCELLANEOUS APPLICATIONS OF... 158

In view of Eq. (7.33), an alternative expression for Eq. (7.35) is as follows:

$$Z_{in} = -[b_1, b_2, \ldots] \begin{bmatrix} V_1 \\ V_2 \\ \vdots \end{bmatrix} \tag{7.36}$$

where

$$V_{q'} = \int_0^d E_{y,I(q')}(x_o, y, z_o) dy \qquad q' = 1, 2, \ldots \tag{7.37}$$

which is similar to Eq. (7.27).

The spectral domain analysis presented above for the rectangular patch antenna is very similar to those reported in publications [26,30]. With some modification, this analysis can be extended in order to deal with more complicated planar radiators such as aperture coupled microstrip antennas and an array of planar radiators [31-34].

Using the above analysis, the variation of input impedance with frequency for a rectangular patch antenna fed by a coaxial probe is shown in Fig. 7.7. In this analysis, the current distribution $J_z(x,z)$ has been approximated as

$$J_z(x,z) = \sin\left[\frac{\pi}{2l}(z+l)\right] \quad -w < x < w, \ -l < z < l \tag{7.38}$$

The experimental verification of the theoretical impedances is also shown in Fig. 7.7.

It should be emphasised yet again that in the technique presented for the analysis of the microstrip patch antenna with feed, Fig. 7.6, the model selected for the feed is approximate [13,28]. This model does not take care of the complicated current distribution at the junction between the patch and the end of the probe, neither does it consider the nonuniform nature of the current along the probe. However, as noted in the literature, the accuracy of such an idealised model is adequately acceptable and no significant error has been observed in the theoretical results as compared with the measurements [28,31].

As regards the computing time associated with the spectral domain analysis of planar radiators, it is very much dependent on how the basis functions for currents or fields are defined. With reference to Eq. (7.15), since double integrals are involved in the expressions for the coefficients of the final matrix, basis functions with slow

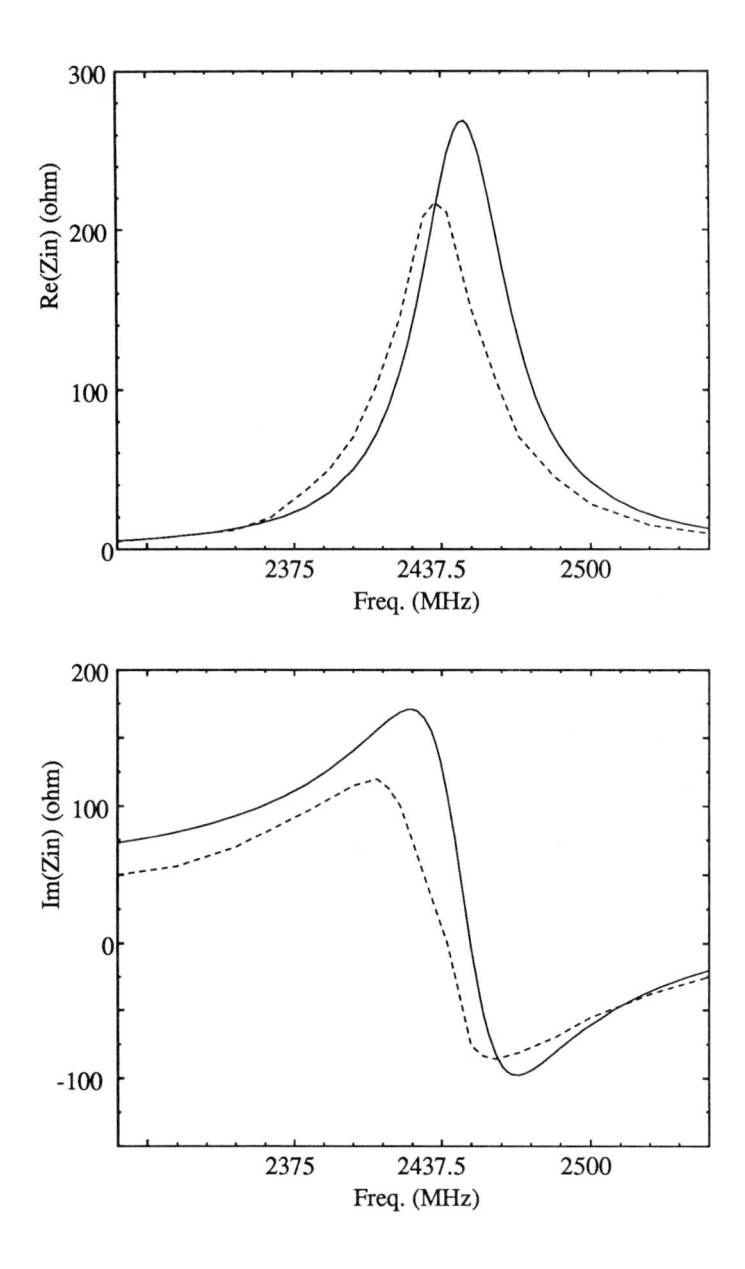

Fig. 7.7 Input impedance of a coaxial fed microstrip antenna (2l=37 mm, 2w=45 mm, d=1.6 mm, x_o=0 mm, z_o=-15 mm and ε_r=2.5); —— spectral domain technique, ---- experiment.

convergence characteristics are not suitable in computations. Examples of appropriate basis functions for rectangular microstrip antennas can be found in references [1,30,33-36]. One such set of trial functions used by Newman et al [35] is given below.

$$J_{z,mn} = \sin\left[\frac{m\pi}{2l}(z+l)\right]\cos\left[\frac{n\pi}{2w}(x+w)\right] \quad \begin{matrix} m = 1,2,...\\ n = 0,1,2,... \end{matrix} \quad (7.39.a)$$

$$J_{x,m'n'} = \cos\left[\frac{m'\pi}{2l}(z+l)\right]\sin\left[\frac{n'\pi}{2w}(x+w)\right] \quad \begin{matrix} m' = 0,1,2,...\\ n' = 1,2,... \end{matrix} \quad (7.39.b)$$

where subscripts mn and $m'n'$ are equivalent to the familiar subscripts q and p respectively. To conclude, it should be mentioned that the computational efficiency of the spectral domain solution of printed radiators may be enhanced using a technique reported in [37].

7.3 Solution of Planar Structures with Periodic Metallisation

Applications of periodic structures in slow wave devices and in filters are well-known [12,38,39]. In microwave integrated circuits, these structures are usually formed by etching a periodic metallisation on the circuit substrate. Examples of some planar periodic structures are shown in Fig. 7.8. These examples and many other similar structures can be analysed using the spectral domain technique. We demonstrate this for the case of a microstrip periodic structure shown in Fig. 7.9.

As in the case of the microstrip line, the spectral domain formulation of the structure shown in Fig. 7.9 is commenced by introducing two scalar potential functions Φ_i and Ψ_i for each dielectric region. These potential functions are functions of x, y and z. Since the propagation of wave in the structure of concern is in the z-direction, the field in the i th region and the associated potential functions (Φ_i and Ψ_i) can be assumed to be

$$E_i = e_i(x,y,z)e^{-j\beta z} \quad (7.40.a)$$
$$H_i = h_i(x,y,z)e^{-j\beta z} \quad (7.40.b)$$
$$\Phi_i = \phi_i(x,y,z)e^{-j\beta z} \quad (7.40.c)$$
$$\Psi_i = \psi_i(x,y,z)e^{-j\beta z} \quad (7.40.d)$$

It is natural to assume that the field in a periodic structure is periodic. Therefore,

CHAP. 7 MISCELLANEOUS APPLICATIONS OF ... 161

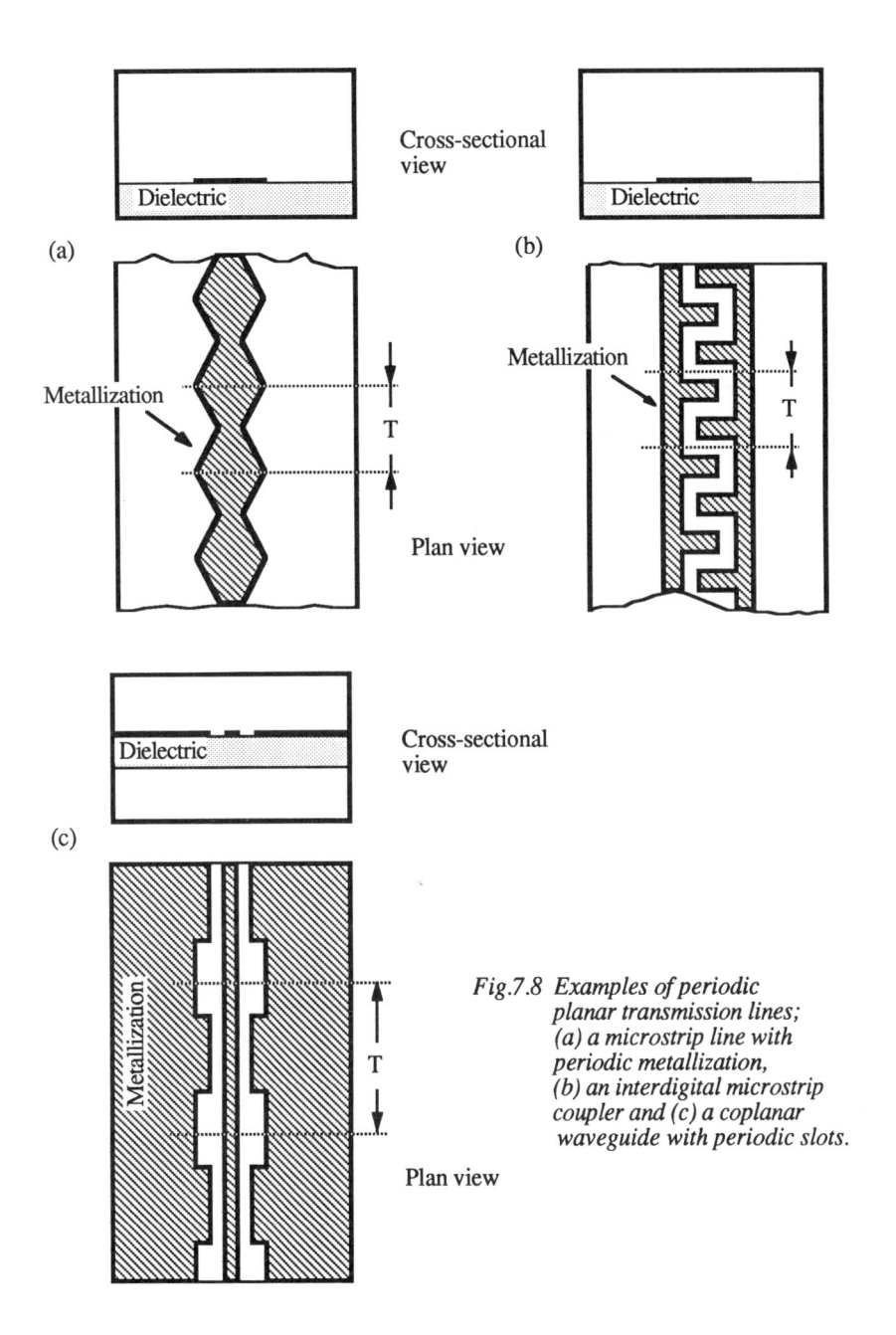

Fig.7.8 Examples of periodic planar transmission lines; (a) a microstrip line with periodic metallization, (b) an interdigital microstrip coupler and (c) a coplanar waveguide with periodic slots.

CHAP. 7 MISCELLANEOUS APPLICATIONS OF ...

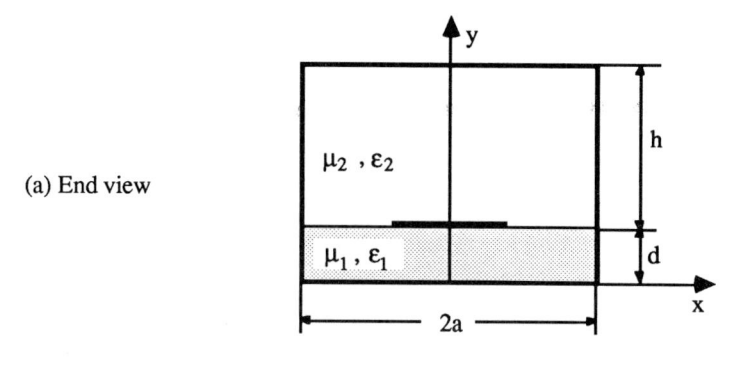

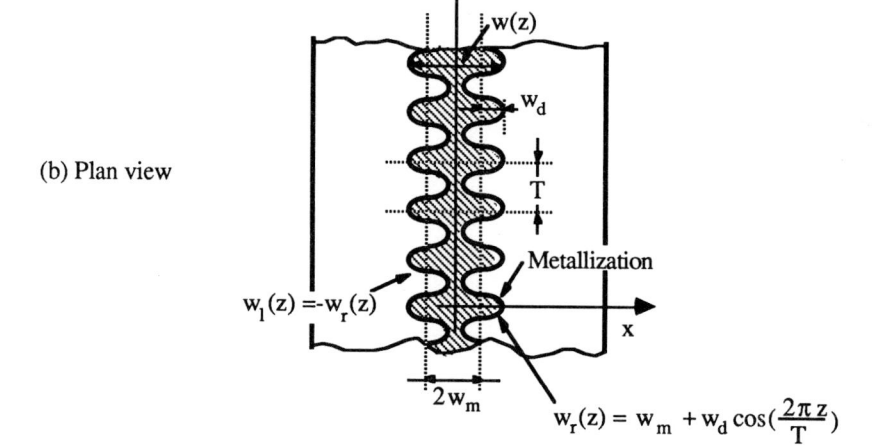

(a) End view

(b) Plan view

$w_r(z) = w_m + w_d \cos(\frac{2\pi z}{T})$

Fig. 7.9 *Microstrip line with sinusoidally varying strip. (From Glandorf and Wolf [40], Copyright © 1987 IEEE, reproduced with permission.)*

functions ϕ_i and ψ_i in the above equations can be expressed in terms of a Fourier series (Floquet's modes) in z [12]:

$$\phi_i = \sum_{k=-\infty}^{+\infty} \tilde{\phi}_i(x,y,\beta_k) e^{-j\beta_k z} \tag{7.41.a}$$

$$\psi_i = \sum_{k=-\infty}^{+\infty} \tilde{\psi}_i(x,y,\beta_k) e^{-j\beta_k z} \tag{7.41.b}$$

CHAP. 7 MISCELLANEOUS APPLICATIONS OF ...

where

$$\beta_k = \frac{2\pi k}{T} \qquad (7.41.c)$$

and T is the periodicity of the structure as shown in Fig. 7.9.

The physical interpretation of Eqs. (7.41) is straightforward; the electromagnetic wave travelling along a periodic structure consists of an infinite number of waves or space harmonics. From Eqs. (7.40) and Eqs. (7.41), it is clear that the phase constant associated with each space harmonic is $\beta + \beta_k$. Now let us consider the potential functions associated with each space harmonic, ie:

$$\bar{\phi}_i(x,y,\beta_k) \text{ and } \bar{\psi}_i(x,y,\beta_k).$$

With reference to the spectral domain solution of the microstrip line in Chapter 3, it is clear that the above potential functions can be expanded in terms of Fourier series as follows:

$$\bar{\phi}_i(x,y,\beta_k) = \sum_{n=-\infty}^{+\infty} \tilde{\phi}_i(\alpha_n,y,\beta_k) e^{-j\alpha_n x} \qquad (7.42.a)$$

$$\bar{\psi}_i(x,y,\beta_k) = \sum_{n=-\infty}^{+\infty} \tilde{\psi}_i(\alpha_n,y,\beta_k) e^{-j\alpha_n x} \qquad (7.42.b)$$

If we substitute Eqs. (7.42) in Eqs. (7.41), the following equations result:

$$\phi_i = \sum_k \sum_n \tilde{\phi}_i(\alpha_n,y,\beta_k) e^{-j(\alpha_n x + \beta_k z)} \qquad (7.43.a)$$

$$\psi_i = \sum_k \sum_n \tilde{\psi}_i(\alpha_n,y,\beta_k) e^{-j(\alpha_n x + \beta_k z)} \qquad (7.43.b)$$

From the above equations, it is not difficult to introduce the following two-dimensional finite Fourier transform:

$$\tilde{u}_i = \frac{1}{4aT} \int_{-a}^{+a} \int_{z_o}^{z_o+T} u_i e^{j(\alpha_n x + \beta_k z)} dx dz \qquad (7.44)$$

where z_o is an arbitrary position along the z-axis. The above transformation is applicable to every component of e_i and h_i (see Eqs. (7.40)) as well as to ϕ_i or to ψ_i. Note that for convenience, z_o in Eq. (7.44) can be selected as $z_o = -T/2$ in which case Eq. (7.44) takes the more familiar form previously introduced by Eq. (3.11.a) for the one-dimensional Fourier transform.

Having established the Fourier transform (7.44), we now proceed to determine the explicit forms of ϕ_i and ψ_i in the spectral domain. With the assumed forms for the potential functions given by Eqs. (7.40.c) and (7.40.d), it is easy to show that ϕ_i and ψ_i are solutions of the wave equation

$$\nabla^2 u_i - 2j\beta \frac{\partial u_i}{\partial z} + (k_i^2 - \beta^2) u_i = 0 \tag{7.45.a}$$

This equation is resulted from Eq. (7.2) when u_i is replaced by $u_i e^{-j\beta z}$ to represent Φ_i and Ψ_i given by Eqs. (7.40.c) and (7.40.d). In the Fourier domain, the above equation is reduced to equation

$$\frac{d^2}{dy^2} \tilde{u}_i - \gamma_{i,nk}^2 \tilde{u}_i = 0 \tag{7.45.b}$$

where

$$\gamma_{i,nk}^2 = \alpha_n^2 + (\beta + \beta_k)^2 - k_i^2 \tag{7.45.c}$$

Comparing Eq. (7.45.b) with Eq. (3.12), it is clear that the Fourier domain expressions of ϕ_i and ψ_i are the same as those given by Eq. (3.15) provided that $\gamma_{i,n}$ is replaced by $\gamma_{i,nk}$. Indeed, from this stage onwards, the spectral formulation of the periodic structure of concern is very similar to that presented for the microstrip line in Chapter 3. Thus, the relation between the spectral components of the current distribution and those of the tangential electric field at the air-dielectric interface can be described as follows:

$$\begin{bmatrix} G_{11}(\alpha_n, \beta_k, \beta) & G_{12}(\alpha_n, \beta_k, \beta) \\ G_{21}(\alpha_n, \beta_k, \beta) & G_{22}(\alpha_n, \beta_k, \beta) \end{bmatrix} \begin{bmatrix} \tilde{J}_x \\ \tilde{J}_z \end{bmatrix} = \begin{bmatrix} \tilde{E}_z \\ \tilde{E}_x \end{bmatrix} \tag{7.46}$$

where $G_{11}(\alpha_n, \beta_k, \beta)$, $G_{12}(\alpha_n, \beta_k, \beta)$ etc. are the same as those given by Eqs. (3.37), with β replaced by $\beta + \beta_k$.

Now let us examine the procedure for converting the above equation to a set of homogeneous linear equations. As mentioned earlier, in a periodic structure, the field is periodic. This entails that J_x and J_z are also periodic along the z-direction. Using

CHAP. 7 MISCELLANEOUS APPLICATIONS OF ...

this information, an appropriate form for expanding J_x and J_z is as follows:

$$J_x = \sum_{p=1}^{\infty} \sum_{l=-\infty}^{+\infty} a_{p,l} J_{x,p}(x,z) e^{-j\frac{2\pi l}{T}z} \quad (7.47.\text{a})$$

$$J_z = \sum_{q=1}^{\infty} \sum_{l=-\infty}^{+\infty} b_{q,l} J_{z,q}(x,z) e^{-j\frac{2\pi l}{T}z} \quad (7.47.\text{b})$$

Note that since factor $e^{-j\beta z}$ is eliminated from both sides of Eq. (7.46), it is not shown in the above expressions. Substituting the Fourier transforms of the above equations in Eq. (7.46) and applying the Galerkin technique to the resulting equation, lead to the desired set of equations. In connection with the microstrip line, the procedure is already described in Chapter 3 and is not repeated here. The nontrivial solution of the set of equations achieved, is obtained by setting the determinant of the coefficients equal to zero. From the resulting equation, the phase constant can be computed.

Although expansions (7.47) are generally acceptable for planar structures with periodic metallizations, basis functions $J_{x,p}(x,z)$ and $J_{z,q}(x,z)$ have to be initially defined according to the geometry of the periodic structure of concern. For instance, in the analysis of the microstrip structure with sinusoidally varying strip that we are concerned with, Fig. 7.9, Glandrof et al [40] employ the following basis functions:

$$\begin{bmatrix} J_{x,p}(x,z) \\ J_{z,p}(x,z) \end{bmatrix} = (1-|\frac{2x}{w(z)}|^2)^{-\frac{1}{2}} \begin{bmatrix} \cos \xi(x,z) & \sin \xi(x,z) \\ -\sin \xi(x,z) & \cos \xi(x,z) \end{bmatrix} \begin{bmatrix} f_{x,p}(x,z) \\ f_{z,p}(x,z) \end{bmatrix} \quad (7.48.\text{a})$$

where

$$f_{x,p}(x,z) = \begin{cases} \sin \dfrac{2p\pi x}{w(z)} & 1 < p < P_z \text{ and } w_r(z) < x < w_l(z) \\ 0 & \text{otherwise} \end{cases} \quad (7.48.\text{b})$$

$$f_{z,p}(x,z) = \begin{cases} \cos \dfrac{2(p-P_z)\pi x}{w(z)} & P_z < p < P \text{ and } w_r(z) < x < w_l(z) \\ 0 & \text{otherwise} \end{cases} \quad (7.48.\text{c})$$

and

$$\xi(x,z) = \frac{2x \arctan [dw_l(z)/dz]}{w(z)} \tag{7.48.d}$$

Note that in the above expressions, P_z is an arbitrary integer such that $P_z < P$ (the first sum in (7.49) is truncated after P terms) and parameters $w_l(z)$, $w_r(z)$ and $w(z)$ are defined in Fig. 7.9. Also note that Glandrof et al [40] employ a vector expansion in order to approximate the current distribution, ie:

$$\begin{bmatrix} J_x \\ J_z \end{bmatrix} = \sum_{p=1}^{\infty} \sum_{l=-\infty}^{+\infty} a_{p,l} \begin{bmatrix} J_{x,p}(x,z) \\ J_{z,p}(x,z) \end{bmatrix} e^{-j\frac{2\pi l}{T}z} \tag{7.49}$$

instead of the scalar expansions (7.47). The reason for the vector representation of basis functions should be sought in the boundary condition which requires the normal component of the current density to vanish at the strip edge. This condition imposes a relation between $a_{p,l}$ and $b_{q,l}$ given in (7.47), making them dependent on each other. This is clearly seen in (7.48.a). In this expression, a term is also included to take care of the edge condition. For further applications of the spectral domain technique to solutions of planar periodic structures, the reader is referred to publications [32,40-43]. Reference [42], which deals with solutions of striplines and finlines with periodic stubs, is particularly interesting for the reported basis functions. In this reference, finlines with periodic stubs, Fig. 7.10, have been analysed using the following basis functions for approximating the tangential components of the electric field at the air-dielectric interface:

$$E_1(x,z) = X(x) e^{-j\beta_0 z} S_1(x,z) \mathbf{i} \tag{7.50.a}$$

$$E_2(x,z) = X(x) e^{-j\beta_{-l} z} S_1(x,z) \mathbf{i} \tag{7.50.b}$$

$$E_3(x,z) = sgn(z) X(x) e^{-j\beta_0 z} S_2(x,z) \mathbf{i} + sgn(x) \cos\left[\frac{\pi x}{2(l+w)}\right] Z(z) S_3(x,z) \mathbf{k} \tag{7.50.c}$$

$$E_4(x,z) = sgn(z) X(x) e^{-j\beta_{-l} z} S_2(x,z) \mathbf{i} + sgn(x) \cos\left[\frac{\pi x}{2(l+w)}\right] Z(z) S_3(x,z) \mathbf{k} \tag{7.50.d}$$

CHAP. 7 MISCELLANEOUS APPLICATIONS OF ... 167

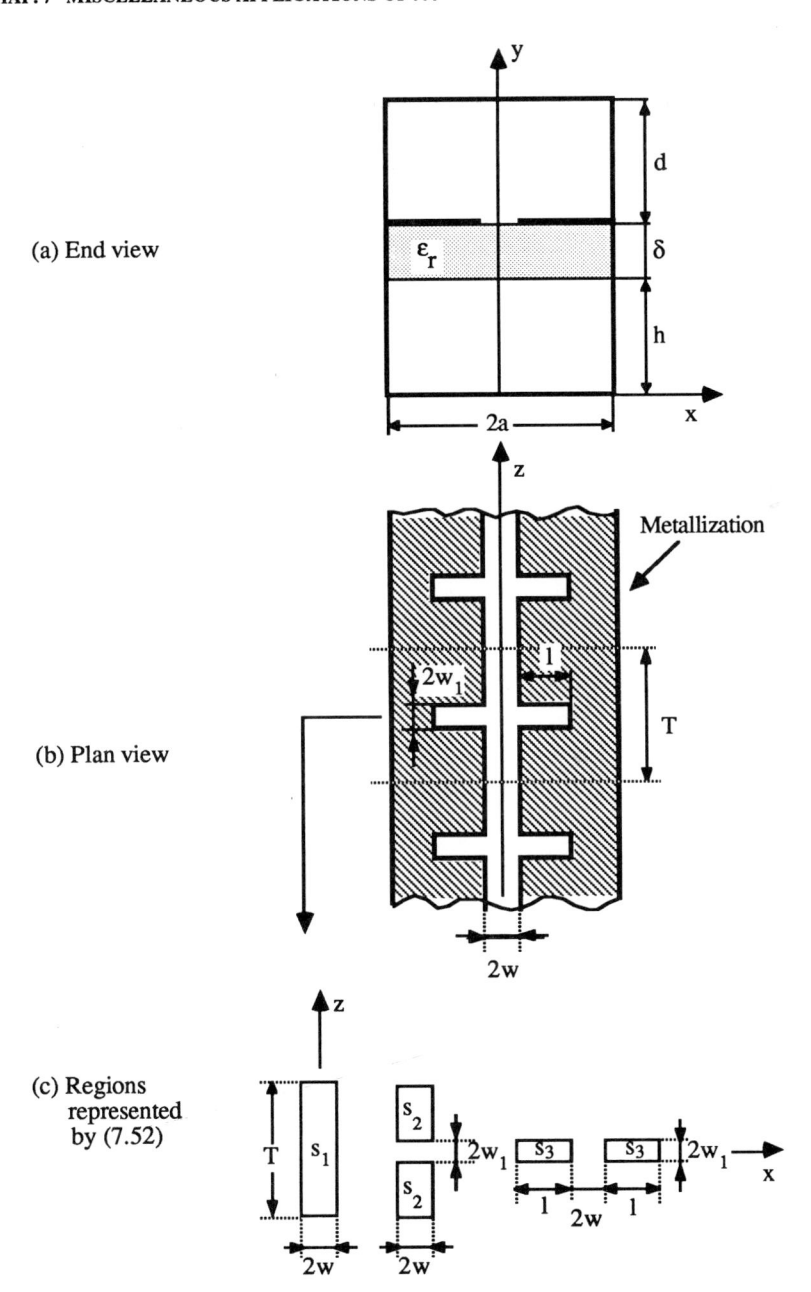

(a) End view

(b) Plan view

(c) Regions represented by (7.52)

Fig. 7.10 Finline with periodic stubs. (From Kitazawa and Mittra [42], Copyright © 1984 IEEE, reproduced with permission.)

where $\beta_{-1} = \beta_0 - 2\pi/T$ (β_0 is the phase constant of the structure) and $X(x)$ and $Z(z)$ denote the x and z variations of the field in the main finline and stubs respectively. The two functions $X(x)$ and $Z(z)$ in Eqs. (7.50) can be expressed in different forms. One simple form is

$$X(x) = C/w \qquad (7.51.a)$$

$$Z(z) = C/w_1 \qquad (7.51.b)$$

where C is a constant. The above choice can be readily justified by referring to Eq. (6.29.a) which represents the approximate form of $E_x(x)$. In the basis functions introduced in Eqs. (7.50), with reference to Fig. 7.10, the auxiliary functions $S_1(x,z)$, $S_2(x,z)$ and $S_3(x,z)$ are defined as follows:

$$S_1(x,z) = \begin{cases} 1 & |x| \leq w \text{ and } |z| \leq T/2 \\ 0 & \text{otherwise} \end{cases} \qquad (7.52.a)$$

$$S_2(x,z) = \begin{cases} 1 & |x| \leq w \text{ and } w_1 \leq |z| \leq T/2 \\ 0 & \text{otherwise} \end{cases} \qquad (7.52.b)$$

$$S_3(x,z) = \begin{cases} 1 & w \leq |x| \leq w+l \text{ and } |z| \leq w_1 \\ 0 & \text{otherwise} \end{cases} \qquad (7.52.c)$$

Note that basis functions (7.50.c) and (7.50.d) are called "junction basis functions" and have also applications in the solution of certain frequency selective surfaces [43].

7.4 Solution of Scattering from Planar Structures

Applications of the spectral domain technique to problems of scattering associated with planar structures are widely recognised in recent years. In this connection, works of Tsao et al [43], Newman et al [35], Uchida et al [44], Pozar [30] and Zhang et al [45] can be regarded as useful starting points for the analysis of scattering problems involving planar structures. In this section, we concentrate on the spectral domain formulation of the scattering of wave by an E-plane rectangular metal stub etched on a dielectric substrate located in a rectangular waveguide [45], Fig. 7.11. It should be mentioned that although the formulation presented below is for a specific

CHAP. 7 MISCELLANEOUS APPLICATIONS OF ...   169

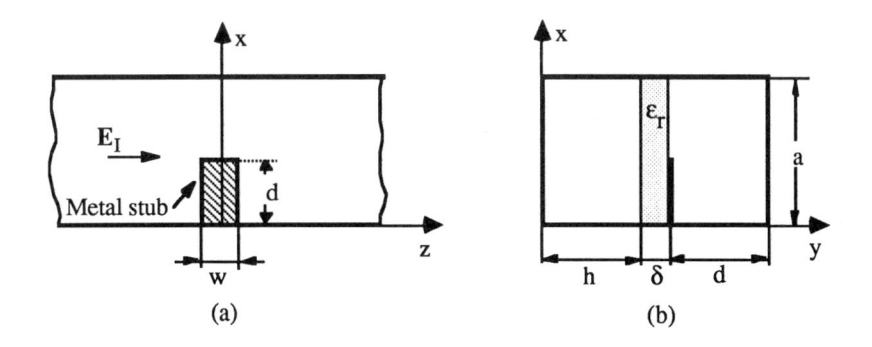

Fig. 7.11 An E-plane rectangular metal stub etched on a dielectric substrate located in a waveguide; (a) longitudinal view and (b) end view. (From Zhang and Itoh [45], Copyright © 1987, reproduced with permission.)

case, it reveals the general procedure in tackling scattering problems by the spectral domain technique.

Consider Fig. 7.11 and assume that the dominant mode is incident on the stub from the left-hand side. This incident field induces a current over the stub, in turn giving rise to a scattered field. The spectral domain relationship between the current on the stub and the tangential components of the electric field at the interface where the stub is present, can be obtained using Eq. (7.20);

$$\begin{bmatrix} G_{11}(\alpha_n,\beta) & G_{12}(\alpha_n,\beta) \\ G_{21}(\alpha_n,\beta) & G_{22}(\alpha_n,\beta) \end{bmatrix} \begin{bmatrix} \tilde{J}_x \\ \tilde{J}_z \end{bmatrix} + \begin{bmatrix} \tilde{E}_{z,I} \\ \tilde{E}_{x,I} \end{bmatrix} = \begin{bmatrix} \tilde{E}_{z,T} \\ \tilde{E}_{x,T} \end{bmatrix} \qquad (7.53)$$

It is clear that the first expression on the left-hand side of the above equation is associated with the scattered field. Note that due to lack of symmetry in the structure of concern in the x-direction, the spectral parameter α_n takes the following form:

$$\alpha_n = \frac{n\pi}{a} \qquad (7.54)$$

which is compatible with Eq. (6.26.a). Also note that the elements of Green's function $G_{11}(\alpha_n,\beta)$, $G_{12}(\alpha_n,\beta)$ etc. are already given in section 6.2. As in the cases of microstrip antennas and microstrip resonators, Eq. (7.53) can be converted into a set of linear equations. This can be achieved by first introducing a set of basis functions for the current distribution:

$$J_x(x,z) = \sum_p a_p J_{x,p}(x,z) \qquad (7.55.a)$$

$$J_z(x,z) = \sum_q b_q J_{z,q}(x,z) \qquad (7.55.b)$$

and then by following the same procedure applied to the cases explained in previous sections. The final equations, which are very similar to Eqs. (3.66) and (3.67), are as follows:

$$\sum_{p=1}^{P} C_{q',p}^{1,1} a_p + \sum_{q=1}^{Q} C_{q',q}^{1,2} b_q = -S_{z,q'} \qquad q' = 1,2,\ldots,Q \qquad (7.56.a)$$

$$\sum_{p=1}^{P} C_{p',p}^{2,1} a_p + \sum_{q=1}^{Q} C_{p',q}^{2,2} b_q = -S_{x,p'} \qquad p' = 1,2,\ldots,P \qquad (7.56.b)$$

where

$$C_{q',p}^{1,1} = \sum_n \int_{-\infty}^{+\infty} G_{11}(\alpha_n,\beta) \tilde{J}_{x,p} \tilde{J}_{z,q'}^* d\beta \qquad (7.57.a)$$

$$C_{q',q}^{1,2} = \sum_n \int_{-\infty}^{+\infty} G_{12}(\alpha_n,\beta) \tilde{J}_{z,q} \tilde{J}_{z,q'}^* d\beta \qquad (7.57.b)$$

$$C_{p',p}^{2,1} = \sum_n \int_{-\infty}^{+\infty} G_{21}(\alpha_n,\beta) \tilde{J}_{x,p} \tilde{J}_{x,p'}^* d\beta \qquad (7.57.c)$$

$$C_{p',q}^{2,2} = \sum_n \int_{-\infty}^{+\infty} G_{22}(\alpha_n,\beta) \tilde{J}_{z,q} \tilde{J}_{x,p'}^* d\beta \qquad (7.57.d)$$

$$S_{z,q'} = \sum_n \int_{-\infty}^{+\infty} \tilde{E}_{z,I} \tilde{J}_{z,q'}^* d\beta \qquad (7.57.e)$$

$$S_{x,p'} = \sum_n \int_{-\infty}^{+\infty} \tilde{E}_{x,I} \tilde{J}_{x,p'}^* d\beta \qquad (7.57.f)$$

Provided that the incident field and trial currents are given, the above coefficients can be evaluated using a computer. Subsequently, coefficients of the current expansions $a_1, a_2, \ldots$ and $b_1, b_2, \ldots$ can be determined by solving the simultaneous equations

(7.56). Accordingly, the evaluation of the components of the equivalent circuits and the scattering parameters of the stub becomes possible. This is achieved by a straightforward modal expansion of the scattered field. Details of the procedure are given in [45].

In practice, since dielectric substrates in E-plane structures are very thin, the incident field can be assumed to be TE_{10}. This assumption allows us to explicitly define $E_{x,I}$ and $E_{z,I}$. As regards the basis functions, J_x and J_z can be either introduced in approximate forms or they can be expanded in series as in Eqs. (7.55). It is clear that the adoption of approximate forms for J_x and J_z contributes to a significant reduction in the computer time whereas a more accurate solution is possible by expanding J_x and J_z in terms of sets of basis functions. In the work reported by Zhang et al [45], basis functions considered for the analysis of the structure shown in Fig. 7.11 are as follows:

$$J_{x,p} = \frac{\cos[(m-1)(\pi z/w + \pi/2)]}{[1-(2z/w)^2]^{\frac{1}{2}}} \times \frac{\sin[(2n-1)(\pi x/2d + \pi/2)]}{[1-(x/d)^2]^{\frac{1}{2}}} \qquad (7.58.\text{a})$$

$$J_{z,q} = \frac{\sin[m'(\pi z/w + \pi/2)]}{[1-(2z/w)^2]^{\frac{1}{2}}} \times \frac{\cos[(2n'-1)(\pi x/2d + \pi/2)]}{[1-(x/d)^2]^{\frac{1}{2}}} \qquad (7.58.\text{b})$$

where m,n,m',n' are integers, and subscripts p and q are defined by a combination of m,n and m',n' respectively. Note that as in the case of the spectral domain analysis of patch antennas, the integrands in Eqs. (7.57) become singular at $\beta's$ associated with dielectric loaded waveguide modes. Hence, as alluded in section 7.2, special care must be exercised in the computation of the coefficients (7.57).

The discontinuity problem addressed above can also be solved using the modal analysis [12] or the least-squares technique [46]. Generally speaking, the application of these techniques requires prior knowledge of the eigenvalues and eigenvectors (phase constants and fields) of the first few modes supported by transmission lines located at the two sides of a discontinuity. This information can be obtained by applying the spectral domain technique to each one of the transmission lines joining at the discontinuity. An instance of such solution can be found in the work of Herald et al [47] where the analysis of several finline discontinuities have been reported. Further examples on the characterisation of nonuniform planar transmission lines can be found in [48]. In that work, analysis of planar tapers using the spectral domain technique in conjunction with the coupled mode theory, is presented.

7.5 Solution of Planar Structures with Lossy and/or Anisotropic Substrates

So far in this book, the spectral domain technique has been examined on the assumption that the layers of materials existing in the construction of a planar structure are lossless and isotropic. Although this assumption is valid for many materials, certain materials used in some planar structures are inherently very lossy and/or strongly anisotropic. For instance, consider the planar slow-wave structures developed on semi-conductor substrates, or monolithic microwave integrated circuits which are based on GaAs [49-51]. It is clear that such planar structures consist basically of at least one lossy substrate. As far as anisotropic materials are concerned, ferrites are extensively used in the development of planar phase shifters, isolators etc. There are also anisotropic dielectric materials which, due to their low costs, are inevitably employed in the design of microwave integrated circuits. For example, Epsilam-10 falls within this category of materials [52].

In this section, we briefly examine the extension of the spectral domain technique to the solution of planar structures consisting of lossy/anisotropic substrates. Let us first start with lossy substrates and show how the losses can be included in the analysis.

Consider Fig. 6.4 and assume that all the dielectric layers are homogeneous, isotropic and lossy. For each layer i, we can define a complex permittivity:

$$\varepsilon_i = \varepsilon'_i - j\varepsilon''_i \tag{7.59}$$

whose imaginary part accounts for the dielectric losses of the layer. Due to the presence of loss in the structure of concern, it is natural to expect that the propagation constants associated with the propagating modes in this structure are complex. As a result, Eqs. (6.1) representing the longitudinal field components in the i th dielectric region can be modified as follows when dielectric losses of that region are taken into account:

$$E_{z,i} = -\frac{k_i^2 + \Gamma^2}{\Gamma} \phi_i e^{-\Gamma z} \tag{7.60.a}$$

$$H_{z,i} = -\frac{k_i^2 + \Gamma^2}{\Gamma} \psi_i e^{-\Gamma z} \tag{7.60.b}$$

CHAP. 7 MISCELLANEOUS APPLICATIONS OF ...

In the above equations Γ denotes the complex propagation constant:

$$\Gamma = \alpha + j\beta \qquad (7.61)$$

Following the same procedure detailed in Chapter 6, it is easy to show that the Fourier domain representation of ϕ_i and ψ_i for the present case are the same as those given by Eqs. (6.2), where $\gamma_{i,n}$ should read as follows:

$$\gamma_{i,n}^2 = \alpha_n^2 - \Gamma^2 - k_i^2 \qquad (7.62)$$

With the new complex expressions (7.59) and (7.62) for ε_i and $\gamma_{i,n}$, it is not difficult to deduce that the characteristic equation of the problem of concern is complex. This equation can be derived by pursuing the same steps explained in Chapter 6. It should be noted that the roots of the characteristic equation are complex (ie: $\Gamma = \alpha + j\beta$). These roots can be sought using a reliable root finding algorithm.

In Fig. 7.12 the results of the spectral domain analysis of the effect of substrate losses on the performance of a microstrip line are depicted. In this case, Muller's algorithm [53] has been employed to compute the roots and hence to determine the normalised wavelengths. Characteristics of certain planar structures involving lossy substrates, evaluated by the spectral domain technique, are reported in [49,50,54].

Let us now study the spectral domain analysis of planar structures consisting of one or several layers of anisotropic materials. Suppose that the *i th* layer in a planar structure is anisotropic. Therefore, the electromagnetic field in that layer satisfies the following form of Maxwell's equations:

$$\nabla \times E_i = -j\omega\mu_i \cdot H_i \qquad (7.63.\text{a})$$

$$\nabla \times H_i = j\omega\varepsilon_i \cdot E_i \qquad (7.63.\text{b})$$

$$\nabla \cdot (\mu_i \cdot H_i) = 0 \qquad (7.63.\text{c})$$

$$\nabla \cdot (\varepsilon_i \cdot E_i) = 0 \qquad (7.63.\text{d})$$

where μ_i and ε_i are second rank tensors representing the permeability and permittivity of the anisotropic layer respectively.

Using the above equations, it is possible to show that the electromagnetic field in the anisotropic layer can be expressed in terms of two potential functions. These potential functions are usually solutions of two coupled second-order differential equations. With reference to Chapter 2, we recall that for an isotropic material, the associated

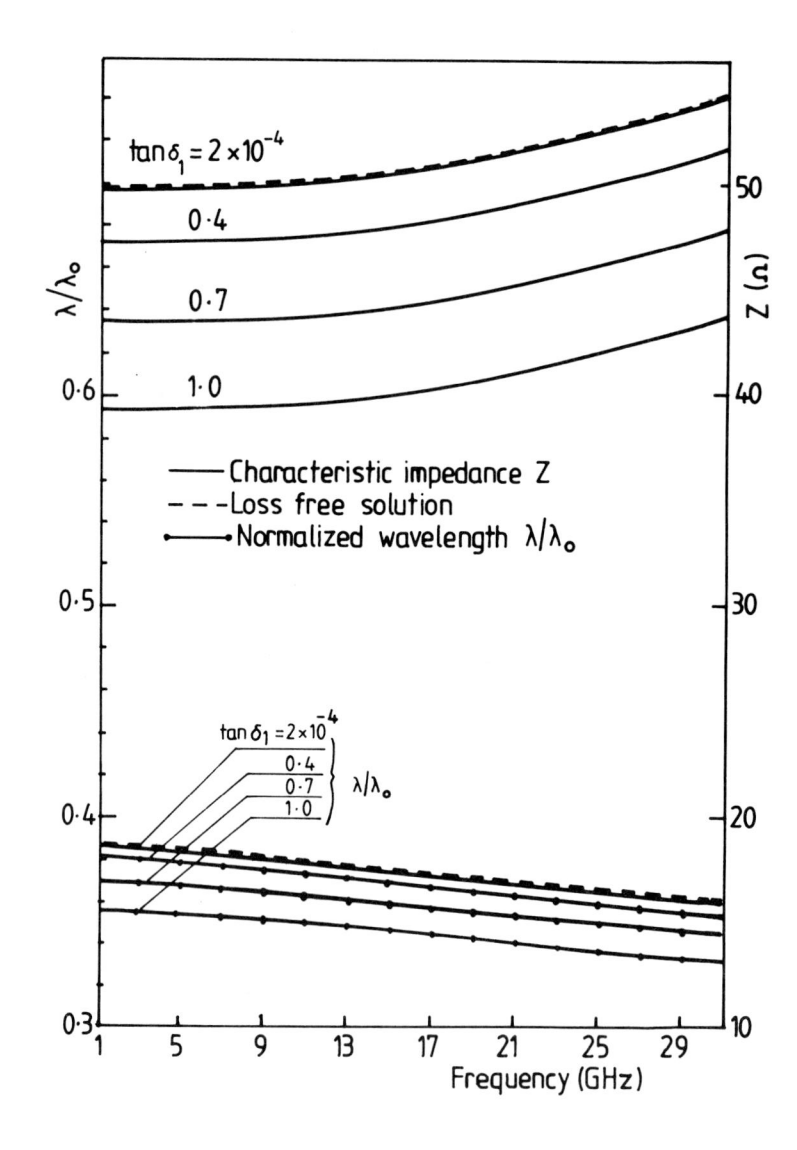

Fig. 7.12 Characteristic impedance and normalized wavelength of the fundamental mode of a microstrip line (2w=d=0.5 mm, h=19.5 mm, 2a=20 mm and $\varepsilon_{r1}=10$) as a function of frequency; computed by the spectral domain technique for different values of dielectric losses. (From Mirshekar-Syahkal [54], Copyright © 1983 IEEE, reproduced with permission.)

differential equations are uncoupled. However, once the Fourier domain potential functions for the anisotropic layer are found (no matter whether these potential functions obey a set of coupled or uncoupled differential equations), the spectral domain formulation of the problem can proceed as that explained in Chapter 6.

As an example, consider the case of a bilateral finline, Fig. 7.13, fabricated on a substrate with permittivity and permeability specified as follows:

$$\varepsilon_1 = \varepsilon_0 \begin{bmatrix} \varepsilon_{xx} & 0 & 0 \\ 0 & \varepsilon_{yy} & 0 \\ 0 & 0 & \varepsilon_{zz} \end{bmatrix} \quad (7.64.a)$$

$$\mu_1 = \mu_0 \quad (7.64.b)$$

Using the above permittivity and permeability in Maxwell's equations (7.63), it is found that the convenient potential functions associated with the substrate are $E_{y,1}$ and $E_{z,1}$. These components satisfy the following coupled differential equations [55]:

$$\frac{\partial^2 E_{y,1}}{\partial x^2} + \frac{\varepsilon_{yy}}{\varepsilon_{xx}} \frac{\partial^2 E_{y,1}}{\partial y^2} + (\varepsilon_{yy} k_0^2 - \beta^2) E_{y,1} + j\beta (1 - \frac{\varepsilon_{zz}}{\varepsilon_{xx}}) \frac{\partial E_{z,1}}{\partial y} = 0 \quad (7.65.a)$$

$$\frac{\partial^2 E_{z,1}}{\partial x^2} + \frac{\partial^2 E_{z,1}}{\partial y^2} + (\varepsilon_{zz} k_0^2 - \beta^2 \frac{\varepsilon_{zz}}{\varepsilon_{xx}}) E_{z,1} + j\beta (1 - \frac{\varepsilon_{yy}}{\varepsilon_{xx}}) \frac{\partial E_{y,1}}{\partial y} = 0 \quad (7.65.b)$$

It is clear that through the terms $\partial E_{z,1}/\partial y$ and $\partial E_{y,1}/\partial y$, the two differential equations in (7.65) are related. Applying the finite Fourier transform (see Eq. (3.11.a)):

$$\tilde{\xi} = \frac{1}{2a} \int_{-a}^{+a} \xi(x,y) e^{j\alpha_n y} dy \quad (7.66)$$

to Eqs. (7.65) leads to the following coupled differential equations in the Fourier domain:

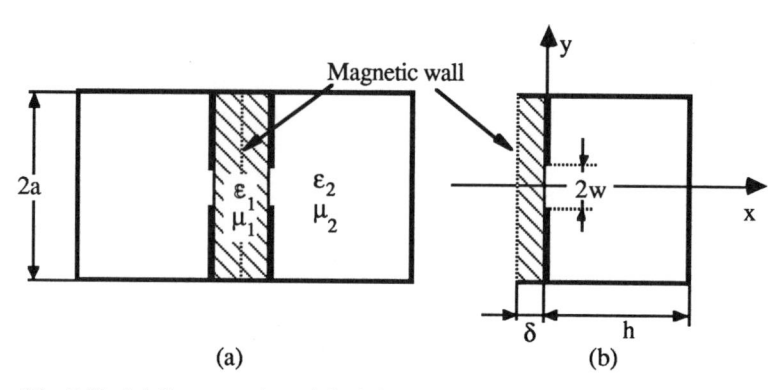

Fig. 7.13 (a) Cross-section of the bilateral finline with an anisotropic dielectric substrate and (b) cross-section used in the analysis [55].

$$\frac{d^2\tilde{E}_{y,1}}{dx^2} + a_1\tilde{E}_{y,1} + a_2\tilde{E}_{z,1} = 0 \qquad (7.67.a)$$

$$\frac{d^2\tilde{E}_{z,1}}{dx^2} + b_1\tilde{E}_{z,1} + b_2\tilde{E}_{y,1} = 0 \qquad (7.67.b)$$

where

$$a_1 = \varepsilon_{yy}k_0^2 - \beta^2 - \frac{\varepsilon_{yy}}{\varepsilon_{xx}}\alpha_n^2 \qquad (7.68.a)$$

$$a_2 = \alpha_n\beta(1 - \frac{\varepsilon_{zz}}{\varepsilon_{xx}}) \qquad (7.68.b)$$

$$b_1 = \varepsilon_{zz}k_0^2 - \beta^2\frac{\varepsilon_{zz}}{\varepsilon_{xx}} - \alpha_n^2 \qquad (7.68.c)$$

$$b_2 = \alpha_n\beta(1 - \frac{\varepsilon_{yy}}{\varepsilon_{xx}}) \qquad (7.68.d)$$

CHAP. 7 MISCELLANEOUS APPLICATIONS OF... 177

From Eqs. (7.67), it is easy to derive the following differential equation:

$$\frac{d^4}{dx^4}\tilde{U}_1 + (a_1+b_1)\frac{d^2}{dx^2}\tilde{U}_1 + (a_1b_1-a_2b_2)\tilde{U}_1 = 0 \qquad (7.69)$$

where U_1 represents $E_{y,1}$ or $E_{z,1}$. The general solution to the above differential equation is straightforward and is available in many calculus text-books. In the present case of interest, Fig. 7.13, the spectral domain forms of $E_{y,1}$ and $E_{z,1}$ are given by [55]:

$$\tilde{E}_{y,1} = s_1 A_{1,n} \cos[\gamma_{1,n}(x+\delta)] + s_2 B_{1,n} \cos[\gamma'_{1,n}(x+\delta)] \qquad (7.70.a)$$

$$\tilde{E}_{z,1} = A_{1,n} \cos[\gamma_{1,n}(x+\delta)] + B_{1,n} \cos[\gamma'_{1,n}(x+\delta)] \qquad (7.70.b)$$

where

$$s_1 = (\gamma_{1,n}^2 - b_1)/b_2 \qquad (7.71.a)$$

$$s_2 = (\gamma'^2_{1,n} - b_1)/b_2 \qquad (7.71.b)$$

$$\gamma_{1,n} = \frac{1}{\sqrt{2}}\left\{(a_1+b_1)+[(a_1-b_1)^2+4a_2b_2]^{\frac{1}{2}}\right\}^{\frac{1}{2}} \qquad (7.71.c)$$

$$\gamma'_{1,n} = \frac{1}{\sqrt{2}}\left\{(a_1+b_1)-[(a_1-b_1)^2+4a_2b_2]^{\frac{1}{2}}\right\}^{\frac{1}{2}} \qquad (7.71.d)$$

For expressions of other field components, the reader may refer to reference [55].

From this stage onwards, the formulation of the problem is similar to that explained in Chapter 3 and Chapter 6. First, we find the necessary Fourier domain field components associated with the air and dielectric region of the structure, Fig. 7.13.b. Then, by using the boundary conditions, we eliminate coefficients $A_{1,n}$ and $B_{1,n}$, together with those coefficients ($A_{2,n}$ and $D_{2,n}$; see Eqs. (6.2.c) and (6.2.d)) associated with the air region in order to obtain the Green's function of the problem. These details of the formulation are reported in [55].

It is interesting to note that when the anisotropic dielectric layer has the following permittivity specification

$$\varepsilon_1 = \varepsilon_0 \begin{bmatrix} \varepsilon_t & 0 & 0 \\ 0 & \varepsilon_{yy} & 0 \\ 0 & 0 & \varepsilon_t \end{bmatrix}, \tag{7.72}$$

we can express the field in terms of two scalar potential functions which are the solutions of two uncoupled wave equations. Examination of Eqs. (7.65) reveals that one of these potential functions can be $E_{y,1}$. This is because the uni-axial axis is normal to the dielectric and conducting layers, and so in view of (7.72), Eq. (7.65.a) reduces to equation

$$\frac{\partial^2 E_{y,1}}{\partial x^2} + \frac{\varepsilon_{yy}}{\varepsilon_t} \frac{\partial^2 E_{y,1}}{\partial y^2} + (\varepsilon_{yy} k_0^2 - \beta^2) E_{y,1} = 0 \tag{7.73}$$

Using Maxwell's equations (7.63), it is possible to establish that the second potential function can be $H_{y,1}$, satisfying equation

$$\frac{\partial^2 H_{y,1}}{\partial x^2} + \frac{\partial^2 H_{y,1}}{\partial y^2} + (\varepsilon_t k_0^2 - \beta^2) H_{y,1} = 0 \tag{7.74}$$

For further reading on the above topic, the reader may refer to publications [52,56,57].

As a further example, let us consider a shielded coplanar waveguide with a ferrite substrate magnetised in the y-direction, Fig. 7.14. For this magnetisation orientation, the permeability tensor of the substrate can be expressed as

$$\mu_2 = \begin{bmatrix} \mu & 0 & -jk \\ 0 & \mu_0 & 0 \\ jk & 0 & \mu \end{bmatrix} \tag{7.75}$$

where expressions for μ and k are given in [58]. Using the above permeability tensor in Maxwell's equations (7.63), it is possible to show that two convenient potential functions for representing the field in the anisotropic layer are $E_{y,2}$ and $H_{y,2}$ [59]. In

CHAP. 7 MISCELLANEOUS APPLICATIONS OF ...   179

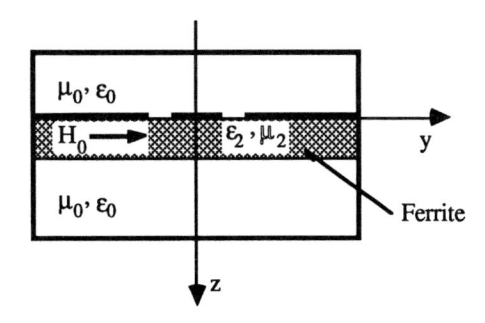

Fig. 7.14 Cross-section of the shielded coplanar waveguide developed on a ferrite substrate. The ferrite is magnetised in the y-direction.

the Fourier domain, these functions satisfy the following differential equations:

$$\frac{d^2\tilde{E}_{y,2}}{dz^2} - (\beta^2 + \alpha_n^2 - \omega^2\varepsilon\frac{\mu^2-k^2}{k})\tilde{E}_{y,2} = j\omega\mu_0\alpha_n\frac{k}{\mu}\tilde{H}_{y,2} \qquad (7.76.\text{a})$$

$$\frac{d^2\tilde{H}_{y,2}}{dz^2} - (\beta^2 + \frac{\mu_0}{\mu}\alpha_n^2 - \omega^2\mu_0\varepsilon)\tilde{H}_{y,2} = -j\omega\varepsilon\alpha_n\frac{k}{\mu}\tilde{E}_{y,2} \qquad (7.76.\text{b})$$

Similar to Eqs. (7.67), the above two equations are coupled. Solutions to these equations are straightforward and can be found in [59]. Therefore, the spectral domain solution of the problem can proceed by following the steps presented in Chapter 6.

References

1. Itoh T., "Analysis of microstrip resonators", *IEEE Trans. Microwave Theory Tech.*, **MTT-22**, pp.946-951, 1974.

2. Sharma A.K. and Baht B., "Spectral domain analysis of interacting microstrip resonant structures", *ibid*, **MTT-31**, pp.681-685, 1983.

3. Kawano K., "Hybrid-mode analysis of coupled microstrip-slot resonators", *ibid*, **MTT-33**, pp.38-43, 1983.

4. Chang T.N. and Wu K.Y., "Transverse modal analysis of edge-coupled

microstrip resonators", *Electron. Lett.*, **Vol. 22**, No. 11, pp.608-609, 1986.

5. Sharma A.K. and Baht B., "Analysis of triangular microstrip resonators", *IEEE Trans. Microwave Theory Tech.*, **MTT-30**, pp.2029-2031, 1982.

6. Sharma A.K., "Spectral domain analysis of an elliptic microstrip ring resonator", *ibid*, **MTT-32**, pp.212-218, 1984.

7. Itoh T. and Mittra R., "A new method for calculating the capacitance of a circular disk for microwave integrated circuits", *ibid*, **MTT-21**, pp.431-432, 1973.

8. Sharma A.K. and Hoefer W.J.R., "Spectral domain analysis of a hexagonal microstrip resonator", *ibid*, **MTT-30**, pp.825-828, 1982.

9. Sharma A.K. and Baht B., "Spectral domain analysis of elliptic microstrip disk resonators", *ibid*, **MTT-28**, pp.573-576, 1980.

10. Chew W.C. and Kong J.A., "Resonance of the axial-symmetric modes in microstrip disk resonators", *J. Math. Phys.*, **Vol. 21**, No. 3, pp.582-591, 1980.

11. Pintzos S.G. and Pregla R., "A simple method for computing the resonant frequencies of microstrip ring resonators", *IEEE Trans. Microwave Theory Tech.*, **MTT-26**, pp.809-813, 1978.

12. Collin R.E., *Field Theory of Guided Waves*, McGraw Hill, New York, 1960.

13. Mosig J.R. and Gardiol F.E., "Dynamical radiation model for microstrip structures", in: *Advances in Electron Physics*, **Vol. 9**, Academic Press, New York, 1982.

14. Wyld H.W., *Mathematical Methods for Physics*, The Benjamin/Cummings Publishing Co. Inc., Menlo Park, California, 1976.

15. Assailly S., Terret C., Daniel J.P., Besnier G., Mosig J. and Roudot B., "Spectral domain approach applied to open resonators: Application to microstrip antennas", *Electron. Lett.*, **Vol. 24**, No. 2, pp.105-106, 1988.

16. Boukamp J. and Jansen R.H., "Spectral domain investigation of surface wave excitation and radiation by microstrip lines and microstrip disk resonators", Proc. 13th European Microwave Conference, pp.721-726, 1983.

17. Uzunoglu N.K., Alexopoulus N.G. and Fikioris J.G., "Radiation properties of microstrip dipoles", *IEEE Trans. Antennas and Propagation*, **AP-27**, pp.853-858, 1979.

18. James J.R., Hall P.S. and Wood C., *Microstrip Antenna Theory and Design*, Peter Peregrinus Ltd., Stevenage, England, 1981.

19. Bahl I.J. and Bhartia P., *Microstrip Antennas*, Artech House, Norwood, MA02062, USA, 1981.

20. Citerne J. and Zieniutycz W., "Spectral domain approach for continuous

spectrum of slot-like transmission lines", *IEEE Trans. Microwave Theory Tech.*, **MTT-33**, pp.817-818, 1985.

21. Hassani H.R. and Mirshekar-Syahkal D., "Full-wave analysis of stacked rectangular microstrip antennas", IEE Conf. Publ. 301, 6th International Conference on Antennas and Propagation, Part 1, pp.369-373, 1989.

22. Wu T.H., Chen K.S. and Peng S.T., "Spectral domain analysis for dielectric antenna loaded with metallic strips", *IEEE,* **MTT-S Digest**, pp.299-301, 1987.

23. Araki K. and Itoh T., "Hankel transform domain analysis of open circular microstrip radiating structures", *IEEE Trans. Antennas and Propagation*, **AP-29**, pp.84-89, 1981.

24. Mittra R. and Kaster R., "A spectral domain approach for computing the radiation characteristic of a leaky-wave antenna for millimeter waves", *ibid*, **AP-29**, pp.652-654, 1981.

25. Itoh T. and Menzel W., "A full-wave analysis method for open microstrip structures", *ibid*, **AP-29**, pp.63-67, 1981.

26. Pozar D.M., "Input impedance and mutual coupling of rectangular microstrip antennas", *ibid*, **AP-30**, pp.1191-1196, 1982.

27. Araki K., Ueda H. and Masayuki T., "Numerical analysis of circular disk microstrip antennas with parasitic elements", *ibid*, **AP-34**, pp.1390-1394, 1986.

28. Newman E.H. and Tulyathan P., "Analysis of microstrip antenna using moment methods", *ibid*, **AP-29**, pp.47-53, 1981.

29. Kong J.A., *Electromagnetic Wave Theory*, John Wiley and Sons, New York, 1986.

30. Pozar D.M., "Radiation and scattering from a microstrip patch on a uniaxial substrate", *IEEE Trans. Antennas and Propagation*, **AP-35**, pp.613-621, 1987.

31. Pozar D.M. and Schaubert D.H., "Analysis of an infinite array of rectangular microstrip patches with idealised probe feeds", *ibid*, **AP-32**, pp.1101-1107, 1984.

32. Pozar D.M. and Schaubert D.H., "Dipole and slot elements and arrays on semi-infinite substrates", *ibid*, **AP-33**, pp.600-607, 1985.

33. Sullivan P.L. and Schaubert D.H., "Analysis of an aperture coupled microstrip antenna", *ibid*, **AP-34**, pp.977-985.

34. Pozar D.M., "A reciprocity method of analysis for printed slot and slot-coupled microstrip antennas", *ibid*, **AP-34**, pp.1439-1446, 1986.

35. Newman E.H. and Forrai D., "Scattering from a microstrip patch", *ibid*, **AP-35**, pp.245-251, 1987.

36. Bailey M.C. and Deshpande M.D., "Integral equation formulation of microstrip

antennas", *ibid*, **AP-30**, pp.651-656, 1982.

37. Pozar D.M., "Improved computational efficiency for the moment method solution of printed dipoles and patches", *Journal of Electromagnetic Society*, **Vol. 3**, No. 3-4, pp.299-309, 1983.

38. Collin R.E., *Foundations for Microwave Engineering*, McGraw Hill, New York, 1966.

39. Ramo S., Whinnery J.R. and Van Duzer T., *Fields and Waves in Communication Electronics*, John Wiley and Sons, New York, 1984.

40. Glandrof F.J. and Wolf I., "A spectral-domain analysis of periodically nonuniform microstrip lines", *IEEE Trans. Microwave Theory Tech.*, **MTT-35**, pp.336-343, 1987.

41. Glandrof F.J. and Wolf I., "A spectral-domain analysis of periodically nonuniform coupled microstrip lines", *ibid*, **MTT-36**, pp.336-343, 1988.

42. Kitazawa T. and Mittra R., "An investigation of striplines and finlines with periodic stubs", *IEEE Trans. Microwave Theory Tech.*, **MTT-32**, pp.684-688, 1984.

43. Tsao C. and Mittra R., "Spectral-domain analysis of frequency selective surfaces comprised of periodic arrays of cross dipoles and Jerusalem crosses", *IEEE Trans. Antennas and Propagation*, **AP-32**, pp.478-486, 1984.

44. Uchida K., Noda T. and Matsunaga T., "Spectral domain analysis of electromagnetic wave scattering by an infinite plane metallic grating", *ibid*, **AP-35**, pp.46-52, 1987.

45. Zhang Q. and Itoh T., "Spectral-domain analysis of scattering from E-plane circuit elements", *IEEE Trans. Microwave Theory Tech.*, **MTT-35**, pp.138-150, 1987.

46. Davies J.B., "A least-squares boundary residual method for the numerical solution of scattering problems", *ibid*, **MTT-21**, pp.99-104, 1973.

47. Herald M., Citerne J., Picon O. and Hana V.F., "Theoretical and experimental investigation of finline discontinuities", *ibid*, **MTT-33**, pp.994-1003, 1985.

48. Mirshekar-Syahkal D. and Davies J.B., "Accurate analysis of tapered planar transmission lines for microwave integrated circuits", *ibid*, **MTT-29**, pp.123-128, 1981.

49. Mu T., Ogawa H. and Itoh T., "Characteristics of multiconductor, asymmetric, slow-wave microstrip transmission lines", *ibid*, **MTT-34**, pp.1471-1477, 1986.

50. Fukuoka C., Shih Y. and Itoh T., "Analysis of slow-wave coplanar waveguide for monolithic integrated circuits", *ibid*, **MTT-31**, pp.567-573, 1983.

51. Finlay H.J., Jansen R.H., Jenkins J.A. and Eddison I.G., "Accurate characterisation and modelling of transmission lines for GaAs MMIC's", *ibid*,

MTT-36, pp.961-966, 1988.

52. Alexopoulos N.G., "Integrated-circuit structures on anisotropic substrates", *ibid*, **MTT-33**, pp.847-881, 1985.

53. Muller D.E., "A method for solving algebraic equations using an automatic computer", *Mathematical Tables and Other Aids to Computations*, **Vol. 10**, pp.208-215, 1956.

54. Mirshekar-Syahkal D., "An accurate determination of dielectric loss effect in monolithic microwave integrated circuits including coupled microstrip lines", *IEEE Trans. Microwave Theory Tech.*, **MTT-31**, pp.950-954, 1983.

55. Yang H. and Alexopoulos N.G., "Uniaxial and biaxial substrate effects on finline characteristics", *ibid*, **MTT-35**, pp.24-29, 1985.

56. Nakatani A. and Alexopoulos N., "Toward a generalised algorithm for the modelling of the dispersive properties of integrated circuit structures on anisotropic substrates", *ibid*, **MTT-33**, pp.1436-1441, 1985.

57. Shalaby A.T.K. and Kumar A., "Dispersion in unilateral finlines on anisotropic substrates", *ibid*, **MTT-35**, pp.448-450, 1987.

58. Lax B. and Button K., *Microwave Ferrites and Ferromagnetics*, McGraw Hill, New York, 1962.

59. El-Sharawy E. and Jackson R.W., "Coplanar waveguide and slot line on magnetic substrates: analysis and experiment", *IEEE Trans. Microwave Theory Tech.*, **MTT-36**, pp.1071-1078, 1988.

Appendix I

FOURIER TRANSFORMS

The theory of Fourier series and Fourier integrals is a well-established subject and its in-depth treatment can be found in the mathematics and physics literature including [1-7]. In this appendix, we briefly review, without presenting any proof, some of the concepts and facts from the theory of the Fourier transforms which have been used throughout this book.

I.1 Expansion of Periodic Functions in terms of Trigonometric Functions (Fourier Series)

The periodic function $f(x)$ with period $2L$, Fig. I.1, can be expressed as a linear combination of trigonometric functions as follows:

$$f(x) = \sum_{n=-\infty}^{+\infty} c_n e^{-j\alpha_n x} \tag{I.1}$$

The above series is called the Fourier series of $f(x)$. In this series coefficients c_n, known as the Fourier coefficients, are given by the following integral

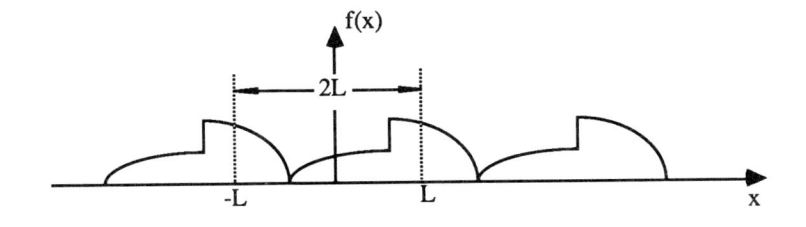

Fig. I.1 A periodic function with period 2L.

APPENDIX I FOURIER TRANSFORMS

$$c_n = \frac{1}{2L} \int_{-L}^{L} f(x) e^{j\alpha_n x} dx \tag{I.2}$$

where

$$\alpha_n = \frac{\pi n}{L} \tag{I.3}$$

From Eq. (I.2), it is clear that the condition for a periodic function to be representable by the Fourier series is the existence of integral (I.2). Assuming that this integral exists, the question of whether the series obtained converges to $f(x)$, can be answered as follows. Sufficient conditions for the convergence of the Fourier series of a periodic function is that the function satisfies Dirichlet's conditions. In the interval $[-L, L]$, a periodic function $f(x)$ is said to satisfy Dirichlet's conditions if
(i) $f(x)$ and $f'(x)$ are piecewise continuous
(ii) $f(x)$ has a finite number of maxima and minima
(iii) the integral of $f(x)$ over the interval $[-L, L]$ is absolutely convergent.

Note that if x is a point of discontinuity for $f(x)$, the Fourier series at this point converges to the following value:

$$\tfrac{1}{2}[f(x+0) + f(x-0)] \tag{I.4}$$

I.2 Fourier Series of Odd and Even Functions

From Eqs. (I.1) and (I.2), it is not difficult to deduce that the Fourier series of an odd function ($f(-x) = -f(x)$) can be represented by

$$f(x) = \sum_{n=1}^{\infty} a_n \sin(\alpha_n x) \tag{I.5}$$

and the Fourier series of an even function ($f(-x) = f(x)$) can be represented by

$$f(x) = \sum_{n=1}^{\infty} b_n \cos(\alpha_n x) \tag{I.6}$$

Note that in Eq. (I.5)

$$a_n = \frac{2}{L} \int_{0}^{L} f(x) \sin(\alpha_n x) \, dx \tag{I.7}$$

and in Eq. (I.6)

$$b_n = \frac{2}{L} \int_0^L f(x)\cos(\alpha_n x)dx \qquad n \neq 0 \qquad (\text{I.8.a})$$

$$b_0 = \frac{1}{L} \int_0^L f(x)dx \qquad (\text{I.8.b})$$

I.3 Half-Range Expansions

In the solution of many engineering and physical problems, there are many circumstances where the expansion of a function $f(x)$ defined on an interval $[0, L]$ is required in terms of sine or cosine functions. In these cases, expansions (I.5) and (I.6) can be used. When $f(x)$ is expanded as in Eq. (I.5), the odd periodic extension of $f(x)$ is generated, Fig. I.2.a, and when it is expanded as in Eq. (I.6), the even periodic extension of $f(x)$ is obtained, Fig. I.2.b. As can be seen from Fig. I.2, these periodic extensions of $f(x)$ are of period $2L$ which is twice the function interval $[0, L]$. For this reason, series (I.5) and (I.6) are referred to as the half-range expansions of $f(x)$.

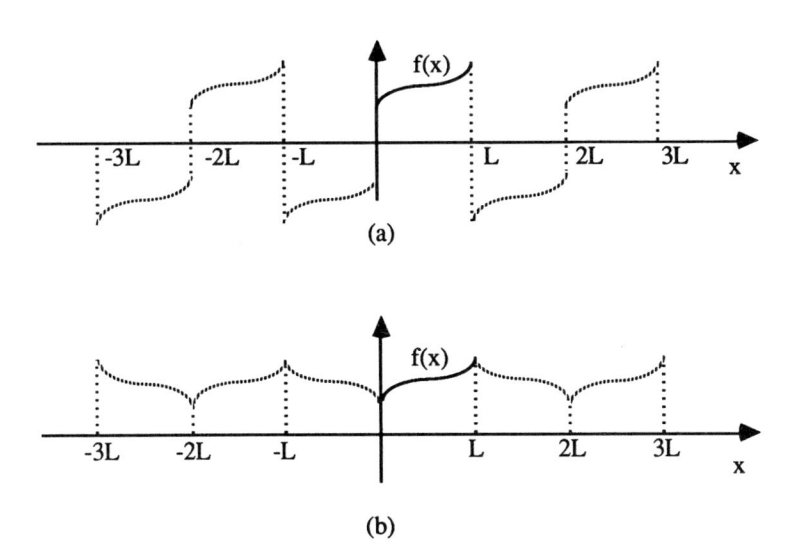

Fig. I.2 Function $f(x)$ and its (a) odd periodic and (b) even periodic extensions.

I.4 Finite Fourier Transform

The finite Fourier transform of a function can be defined in two different forms of sine and cosine transforms [1]. In fact, Eqs. (I.7) and (I.8) represent the finite Fourier sine transform and the finite Fourier cosine transform of function $f(x)$, respectively. Note that $f(x)$ is assumed to be defined over the interval $[0, L]$.

The definition presented in this section for the finite Fourier transform of a function $f(x)$ defined over the interval $[0, L]$ is more general and covers the above transformations simultaneously. The definition is based on the fact that the Fourier sine and cosine expansions (half-range expansions) of a function $f(x)$ can be combined into one expansion defined over the extended interval $[-L, L]$; ie:

$$f(x) = \sum_{n=-\infty}^{+\infty} c_n e^{-j\alpha_n x} \qquad (I.9)$$

where

$$c_n = \frac{1}{2L} \int_{-L}^{+L} f(x) e^{j\alpha_n x} dx \qquad (I.10)$$

It is clear that (I.9) and (I.10) can be converted to (I.5) and (I.7) if $f(x)$ is assumed an odd function, and to (I.6) and (I.8) if $f(x)$ is assumed an even function. Note that although $f(x)$ is defined over the interval $[0, L]$ only, we artificially extend its range to $[-L, L]$.

In view of the above discussion, the finite Fourier transform of function $f(x)$ can now be defined generally by

$$\tilde{f}(\alpha_n) = \frac{1}{2L} \int_{-L}^{+L} f(x) e^{j\alpha_n x} dx \qquad (I.11)$$

In short notation (I.11) is shown by

$$\tilde{f} = F(f(x)) \qquad (I.12)$$

Using the above definition, Eq. (I.9) can be written in the following form

$$f(x) = \sum_{n=-\infty}^{+\infty} \tilde{f}(\alpha_n) e^{-j\alpha_n x} \qquad (I.13)$$

which is the inverse Fourier transform of the finite Fourier transform defined by Eq. (I.11).

APPENDIX I FOURIER TRANSFORMS   189

I.5 Fourier Integral and Fourier Transform

In section I.1, the expansion of a periodic function $f(x)$ in terms of trigonometric functions is considered. When $f(x)$ is not periodic and $L \to \infty$, it is possible to show that the Fourier series (I.1) is replaced by the integral

$$f(x) = \int_{-\infty}^{+\infty} \tilde{f}(\alpha) e^{-j\alpha x} d\alpha \qquad (I.14)$$

where

$$\tilde{f}(\alpha) = \frac{1}{2\pi} \int_{-\infty}^{+\infty} f(x) e^{j\alpha x} dx \qquad (I.15)$$

Eq. (I.14) is called the Fourier integral expansion of $f(x)$.

By definition, Eq. (I.15) is the Fourier transform of $f(x)$ and its pair, Eq. (I.14), is the inverse Fourier transform. Note that this definition is consistent with the definition of the finite Fourier transform persented in the previous section. Also, note that notation (I.12) is used for representing (I.15).

I.6 Fourier Transforms of the Derivatives of a Function

The application of the Fourier transform in the solutions of differential equations usually requires the knowledge of Fourier transforms of the derivatives of a function. Using notation (I.12) for the Fourier transform of $f(x)$, it is easy to show that the Fourier transform of the m th derivative of $f(x)$ is

$$F\left(\frac{d^m f(x)}{dx^m}\right) = (-i\alpha)^m \tilde{f} \qquad (I.16)$$

provided that

$$f, \frac{df}{dx}, \ldots, \frac{d^{m-1}f}{dx^{m-1}} \to 0 \text{ as } x \to \pm\infty \qquad (I.17)$$

When F represents the finite Fourier transform, Eq. (I.16) reads as follows:

$$F\left(\frac{d^m f(x)}{dx^m}\right) = (-i\alpha_n)^m \tilde{f} \qquad (I.18)$$

For this equation to be held true, we should generally have

APPENDIX I FOURIER TRANSFORMS

$$f, \frac{df}{dx}, \ldots, \frac{d^{m-1}f}{dx^{m-1}} = 0 \quad at \quad x = \pm L \tag{I.19}$$

I.7 Parseval's Identities

Without going into details of its proof, Parseval's identity associated with the finite Fourier transform is

$$\frac{1}{2L} \int_{-L}^{+L} f(x)g^*(x)dx = \sum_{n=-\infty}^{+\infty} \tilde{f}(\alpha_n)\tilde{g}^*(\alpha_n) \tag{I.20}$$

and that associated with the Fourier transform is

$$\frac{1}{2\pi} \int_{-\infty}^{+\infty} f(x)g^*(x)dx = \int_{-\infty}^{+\infty} \tilde{f}(\alpha)\tilde{g}^*(\alpha)d\alpha \tag{I.21}$$

where (*) denotes the complex conjugate.

References

1. Sneddon I.N., *Fourier Transforms*, McGraw Hill, New York, 1951.

2. Carslaw H.S., *Introduction to the Theory of Fourier's Series and Integrals*, Dover Publications, New York, 1930.

3. Arsac J., *Fourier Transforms and the Theory of Distribution*, Prentice-Hall, Englewood Cliffs, N.J., 1966.

4. Lighthill M.J., *An Introduction to Fourier Analysis and Generalised Functions*, Cambridge University Press, Cambridge, 1958.

5. Lanczos C., *Discourse on Fourier Series*, Oliver and Boyd, Edinburgh and London, 1966.

6. Walker P.L., *The Theory of Fourier Series and Integrals*, John Wiley and Sons, New York, 1986.

7. Churchill R.V., *Fourier Series and Boundary Value Problems*, McGraw Hill, New York, 1963.

Appendix II

LINEAR INTEGRAL EQUATIONS

This appendix is devoted to a brief discussion of types of linear integral equations. It also examines the relation between Green's function and integral equations.

II.1 Classification

An equation of the form

$$g(\xi) + \lambda \int_\tau \ldots \int K(\xi, \eta) f(\eta) d\gamma_\eta = f(\xi) \qquad (\text{II.1})$$

where $g(\xi)$ is a known function and $f(\xi)$ is the function to be determined, is called Fredholm's integral equation of the second kind. In this equation, the numerical parameter λ, the fixed integration domain τ and the kernel of the integral equation $K(\xi, \eta)$, are all defined parameters. The one-dimensional version of the integral equation (II.1) is

$$g(x) + \lambda \int_a^b K(x, s) f(s) ds = f(x) \qquad (\text{II.2})$$

where the integration is over the fixed interval $[a, b]$.

Using the same notations as in Eq. (II.1), Fredholm's integral equation of the first kind is defined by the equation

$$\int_\tau \ldots \int K(\xi, \eta) f(\eta) d\gamma_\eta = g(\xi) \qquad (\text{II.3})$$

APPENDIX II LINEAR INTEGRAL EQUATIONS   192

When the integration domain τ is a variable, Eq. (II.1) and Eq. (II.3) are called the Volterra integral equation of the second and the first kind respectively. An example of the Volterra equation of the second kind in the one-dimensional case is the equation

$$g(x) + \lambda \int_a^x K(x,s)f(s)ds = f(x) \qquad (II.4)$$

In Eq. (II.1), when the inhomogeneous term $g(\xi) = 0$, an "eigenvalue integral equation" results

$$\int_\tau \cdots \int K(\xi, \eta) f(\eta) d\gamma_\eta = \frac{1}{\lambda} f(\xi) \qquad (II.5)$$

For the sake of convenience, the following discussions are concentrated around one-dimensional integral equations. Concepts to be introduced are, however, generally applicable to multidimensional integral equations.

II.2 Types of Kernels

The kernel of an integral equation $K(x,s)$, is called degenerate (separable) if it can be written in one of the following forms

$$K(x,s) = u(x)v(s) \qquad (II.6.a)$$

$$K(x,s) = \sum_i u_i(x)v_i(s) \qquad (II.6.b)$$

Integral equations with degenerate kernels are usually easier to solve. For example, consider Fredholm's integral equation of the second kind, Eq. (II.2), and assume that its kernel is separable, as given by Eq. (II.6.b). Hence, Eq. (II.2) can be written in the following form:

$$g(x) + \lambda \sum_i u_i(x) C_i = f(x) \qquad (II.7.a)$$

where

$$C_i = \int_a^b v_i(s)f(s)ds \qquad (II.7.b)$$

Now by substituting for $f(s)$ in Eq. (II.7.b) from Eq. (II.7.a), the following system of equations results:

APPENDIX II LINEAR INTEGRAL EQUATIONS

$$C_i - \lambda \sum_j k_{ij} C_j = r_i \qquad i = 1, 2, \ldots \qquad (\text{II.8.a})$$

where

$$k_{ij} = \int_a^b v_i(s) u_j(s) ds \qquad (\text{II.8.b})$$

$$r_i = \int_a^b v_i(s) g(s) ds \qquad (\text{II.8.c})$$

From Eq. (II.8.a), coefficients C_i and subsequently $f(x)$ from Eq. (II.7.a) can be determined.

An important application of separable kernels can be found in the solution of integral equations with nonseparable kernels. Usually a nonseparable kernel is approximated by a suitable separable kernel [1,2]. Consequently the approximate solution of the integral equation whose kernel is approximated, can be determined using the procedure outlined above.

The kernel of an integral equation is said to be symmetric if it satisfies the condition

$$K(x,s) = K(s,x) \qquad (\text{II.9})$$

and is said to be Hermitian if it satisfies the condition

$$K(x,s) = K^*(s,x) \qquad (\text{II.10})$$

where superscript (*) denotes the complex conjugate. A kernel is called positive definite if for any piecewise continuous function $f(x)$, the following relation holds:

$$\int_a^b \int_a^b K(x,s) f(s) f(x) dx ds > 0 \qquad (\text{II.11})$$

For the study of properties of integral equations with the above defined kernels, the reader may refer to [3].

The last category of kernels to be introduced here concerns kernels which are functions of $(x-s)$. Solutions to some integral equations involving this type of kernel can be achieved using the Fourier transform technique. For instance, consider the following integral equation:

$$g(x) + \lambda \int_{-\infty}^{+\infty} K(x-s) f(s) ds = f(x) \qquad (\text{II.12})$$

The application of the Fourier transform technique, presented in Appendix I, allows us

APPENDIX II LINEAR INTEGRAL EQUATIONS   194

to write the above equation in the following form [4]:

$$\tilde{g}(\alpha) + 2\pi\lambda\tilde{K}(\alpha)\tilde{f}(\alpha) = \tilde{f}(\alpha) \tag{II.13}$$

where (~) denotes the Fourier transform operator; (see Eqs. (I.12) and I.15)). From the above equation, we have

$$\tilde{f}(\alpha) = \frac{\tilde{g}(\alpha)}{1-2\pi\lambda\tilde{K}(\alpha)} \tag{II.14}$$

By applying the inverse Fourier transform, Eq. (I.14), to the above equation, $f(x)$ is determined.

II.3 Green's Function and its Role in Integral Equations

In the mathematical modelling of many physical phenomena involving sources, the governing differential equations may be replaced by integral equations using Green's functions. For example, consider the following well-known differential equation:

$$\frac{d}{dx}[p(x)y'(x)] + q(x)y(x) = f(x) \tag{II.15.a}$$

subject to boundary conditions

$$y(a) = y(b) = 0 \tag{II.15.b}$$

where $a \leq x \leq b$. From Eq. (II.15.a), it is clear that $f(x)$ is the source function.

Now let us replace the source $f(x)$ in Eq. (II.15.a) by a point source $\delta(x-s)$ which is located at $x=s$. The solution to the resulting differential equation is called Green's function of the problem and it is a function of x and s. Mathematically speaking, Green's function $G(x,s)$ associated with the present problem satisfies the following equations:

$$\frac{d}{dx}[p(x)G_x'(x,s)] + q(x)G(x,s) = \delta(x-s) \tag{II.16.a}$$

$$G(a,s) = G(b,s) = 0 \tag{II.16.b}$$

Using the superposition principle, it is not difficult to show that the solution to the differential equation (II.15.a) is

$$y(x) = \int_a^b G(x,s)f(s)ds \tag{II.17}$$

From the above brief discussion, it is clear that Green's function $G(x,s)$ is simply the

APPENDIX II LINEAR INTEGRAL EQUATIONS

solution of the homogeneous equation

$$\frac{d}{dx}[p(x)G'_x(x,s)] + q(x)G(x,s) = 0 \tag{II.18}$$

for all points in the interval $[a, b]$ with the exception of the point $x=s$. At this point $G'_x(x,s)$ is discontinuous. It can be easily shown that the magnitude of this discontinuity is [5]

$$\lim_{\varepsilon \to 0} G'_x(s+\varepsilon, x) - G'_x(s-\varepsilon, x) \to -\frac{1}{p(s)} \tag{II.19}$$

Now let us assume that the source function $f(x)$ is not known, but its effect $y(x)$ is given. In this case, Eq. (II.17) corresponds to a Fredholm's integral equation of the first kind whose kernel is $G(x,s)$. In connection with the above discussion let us consider the following example.

Assume that a perfect conducting strip of negligible thickness and of infinite length is at the electrostatic potential V_0, Fig. II.1. Let $\sigma(x)$ represent the charge density on the strip. For this example the Green's function $G(x,y,s)$ satisfies the differential equation [6]

$$\nabla^2 G(x,y,s) = -\frac{1}{\varepsilon}\delta(x-s)\delta(y) \tag{II.20}$$

Solving the above equation leads to the following expression for $G(x,y,s)$ [6,7]

$$G(x,y,s) = \frac{1}{2\pi\varepsilon} ln[(x-s)^2 + y^2]^{-\frac{1}{2}} \tag{II.21}$$

From the above equation, it is clear that the obtained Green's function represents the potential due to a line charge at the coordinate $(s,0)$. Thus, the potential due to the

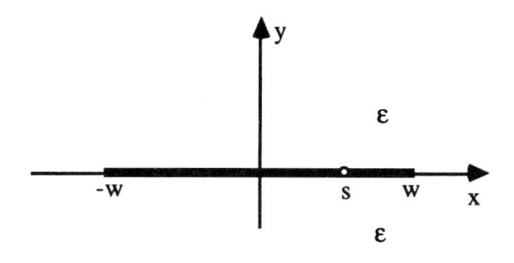

Fig. II.1 Cross-section of a conducting strip of infinite length and of negligible thickness.

charged strip is

$$V(x,y) = \int_{-w}^{+w} G(x,y,s)\sigma(s)ds \tag{II.22}$$

Since $\sigma(x)$ is not known, Eq. (II.22) cannot reveal any information. However, we know that on the strip

$$V(x,0) = V_0 \qquad |x|<w \tag{II.23}$$

Hence, Eq. (II.22) on the strip reads as follows:

$$\frac{1}{2\pi\varepsilon} \int_{-w}^{+w} ln|x-s|^{-1}\sigma(s)ds = V_0 \qquad |x|<w \tag{II.24}$$

The above equation which is a Fredholm's integral equation of the first kind, can be solved using one of the available standard techniques [1,2].

The above example shows how a differential equation involving a source with unknown distribution can be turned into an integral equation whose solution gives the source distribution. The kernel of the integral equation is the Green's function representing the solution of the differential equation for a point source of unit strength.

For further study of integral equations and Green's functions, references [8-14] are recommended.

References

1. Kantorovich L.V. and Krylov V.I., *Approximate Methods of Higher Analysis*, Interscience Publishers, New York, 1964.

2. Todd J. (ed.), *Survey of Numerical Analysis*, McGraw Hill, New York, 1962.

3. Arsenin V. Ya., *Basic Equations and Special Functions of Mathematical Physics*, Iliffe Book Ltd., London, 1968.

4. Wyld H.W., *Mathematical Methods for Physics*, The Benjamin/Cummings Publishing Company, Menlo Park, California, 1976.

5. Dettman J.W., *Mathematical Methods in Physics and Engineering*, McGraw Hill, New York, 1962.

6. Collin R.E., *Field Theory of Guided Waves*, McGraw Hill, New York, 1960.

7. Van Bladel J., *Electromagnetic Fields*, McGraw Hill, New York, 1964.

8. Green C.D., *Integral Equation Methods*, Thomas Nelson and Sons Ltd., London, 1969.

9. Stakgold I., *Green's Functions and Boundary Value Problems*, John Wiley and Sons, New York, 1979.

10. Jaswon M.A. and Symm G.T., *Integral Equation Methods in Potential Theory and Elàstostatics*, Academic Press, London, 1977.

11. Cushing J.T., *Applied Analytical Mathematics for Physical Scientists*, John Wiley and Sons, New York, 1975.

12. Morse P.M. and Feshbach H., *Methods of Theoretical Physics*, McGraw Hill, New York, 1953.

13. Sommerfeld A., *Partial Differential Equations in Physics*, Academic Press, New York, 1949.

14. Courant R. and Hilbert D., *Methods of Mathematical Physics*, Interscience Publishers, New York, 1953.

Appendix III

COMPUTER PROGRAMS

Computer programs which are discussed in this appendix are based on the spectral domain technique. All these programs are written in FORTRAN. The first program, which is very versatile, computes propagation constants of planar transmission lines with cross-sectional geometries as shown in Fig. III.1. Therefore, shielded microstrip lines with single or coupled strips, shielded slot lines with single or coupled slots, shielded coplanar waveguides and unilateral finlines with single or coupled gaps can all be analysed easily by the program. Of course, the named transmission lines are just some examples of a wide range of transmission lines which can be analysed by the program. It should be mentioned that the first program is meant to introduce the beginner to spectral domain programming. Subroutines of this program may find applications in the development of computer programs of other planar transmission lines. Note specially that the program employs a subroutine for generating the spectral domain Green's function associated with three-layer planar transmission lines introduced earlier in section 6.2. This subroutine can be employed in the spectral domain computations of other three-layer transmission lines with strip or slot distributions different from those specified in Fig. III.1.

The programs following the first computer program are subroutines developed originally for the computation of the spectral domain Green's functions of bilateral finlines, antipodal finlines and shielded coupled strip-slot lines [1-3]. Of course, applications of these subprograms are not limited to the above named transmission lines. In fact, the presented subroutines can be employed in the spectral domain computation of any transmission line with a cross-section compatible with one of the general cross-sections shown in Fig. III.2. For further information, the reader may refer to sections 6.2, 6.2.3 and 6.3.1 where the above aspects are covered.

APPENDIX III COMPUTER PROGRAMS

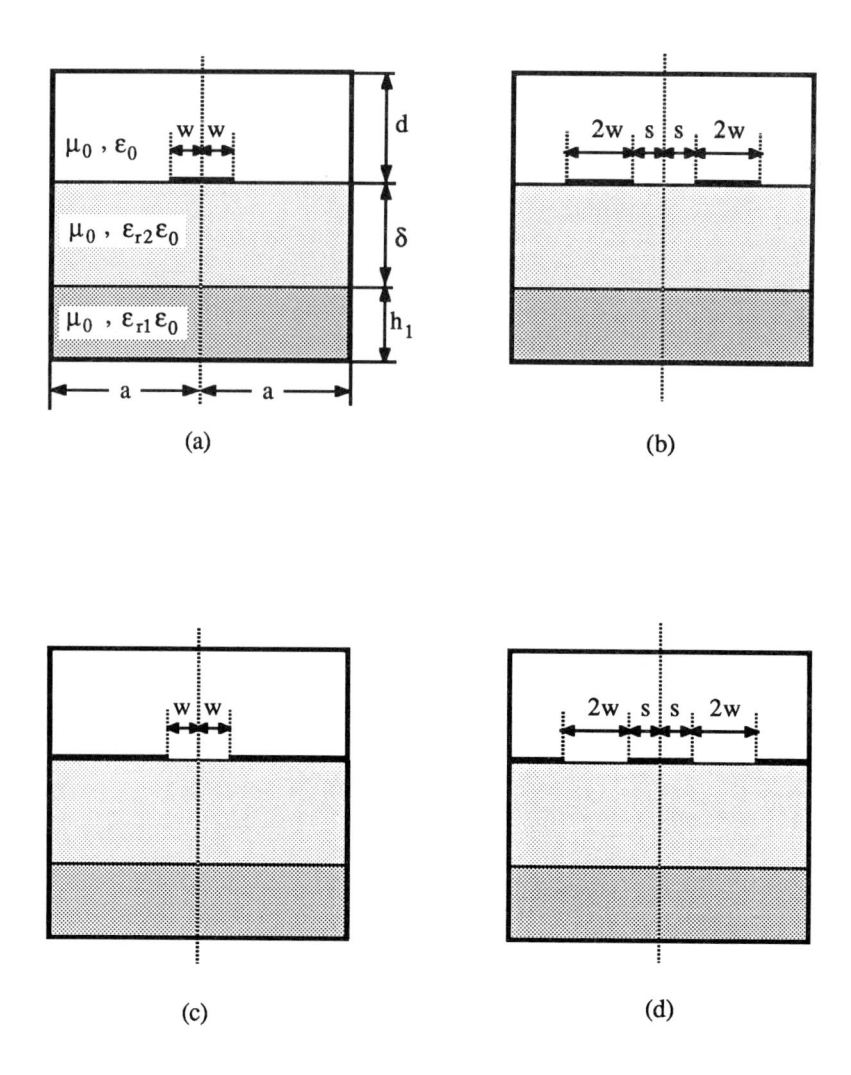

Fig. III.1 Cross-sections of planar transmission lines whose phase constants can be computed by the computer program MRST1.

III.1 Computer Program MRST1

The computer program **MRST1** is listed in section III.5. This program computes dispersion characteristics of planar transmission lines whose cross-sections are

APPENDIX III COMPUTER PROGRAMS

compatible with the general cross-sections shown in Fig. III.1. For example, consider the conventional microstrip line, Fig. III.3.a and the coupled unilateral finline, Fig. III.3.b. The cross-section of the former structure corresponds to that in Fig. III.1.a when $h_1 = 0$, whereas the latter structure has a cross-section compatible with Fig. III.1.d when $\varepsilon_{r1} = 1$ and $h_1 = d$.

The flow-chart for **MRST1** is shown in Fig. III.4. Although the way that the program operates is highlighted by the flow-chart, for the benefit of the reader, brief

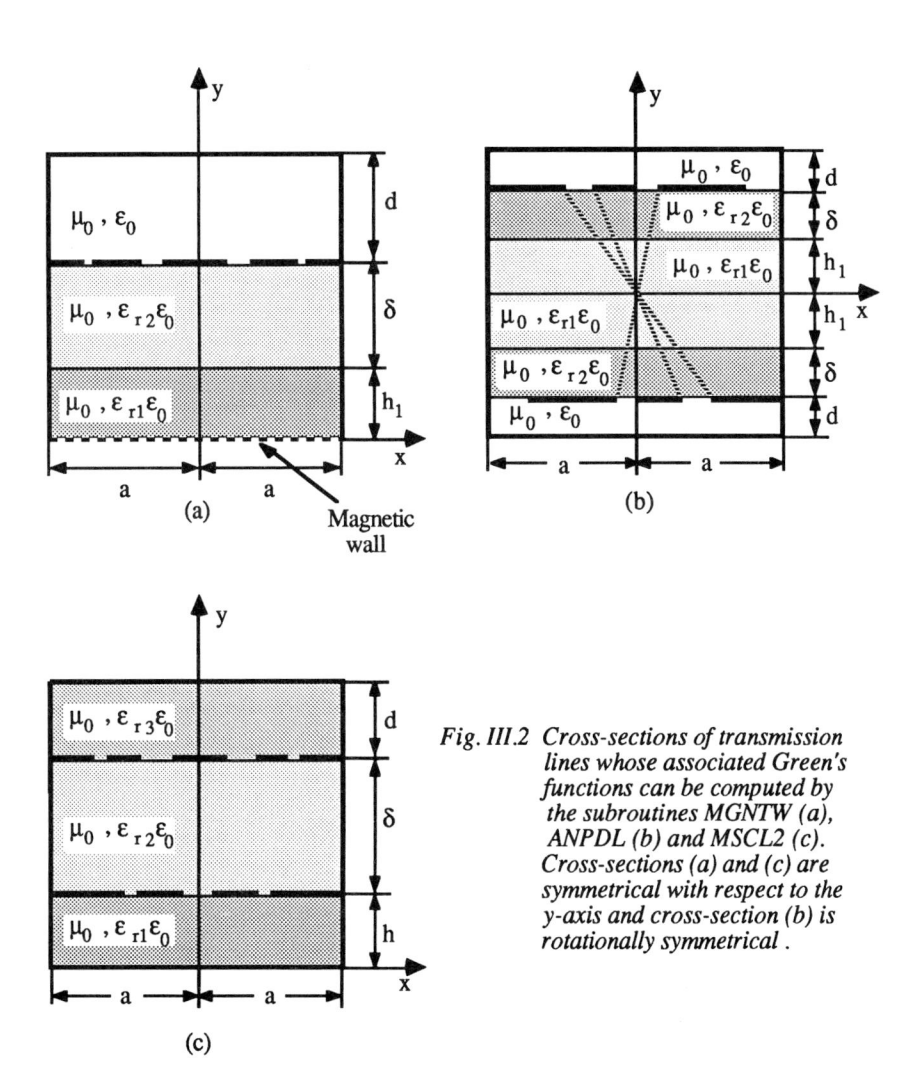

Fig. III.2 Cross-sections of transmission lines whose associated Green's functions can be computed by the subroutines MGNTW (a), ANPDL (b) and MSCL2 (c). Cross-sections (a) and (c) are symmetrical with respect to the y-axis and cross-section (b) is rotationally symmetrical.

APPENDIX III COMPUTER PROGRAMS 202

descriptions of various parts of the program are given below.

MAIN PROGRAM

This part of the program performs various tasks including:

1. Reading the data
2. Calling various subroutines in order to generate the final matrix and to compute its determinant
3. Computing the roots of the determinant
4. Writing the results and messages

The **MAIN PROGRAM** initially reads in the data in the order and by the format specified in the program (see the comments in the program listing). With reference to Fig. III.1 the required data associated with the geometry and material properties are as follows:

Data	Computer Equivalent
ε_{r1}	ER1
ε_{r2}	ER2
h_1	H1
δ	TH
d	D
a	A
w	W
s	S

Note the differences in the definitions of **W** and **S** for different cross-sections shown in Fig. III.1. In addition to the above set, further data required by the program are as follows. With reference to the program, **ISS** and **ICODE** specify the structure and the mode for which the program is run. For structures classified by Fig. III.1.a-b, **ISS=6** and for those categorised by Fig. III.1.c-d, **ISS=5**. For structures with coupled strips or slots, **ICODE** is assigned **+1** for the even mode and **-1** for the odd mode. When the structure consists of a single strip or a single slot, **ICODE** must be

APPENDIX III COMPUTER PROGRAMS

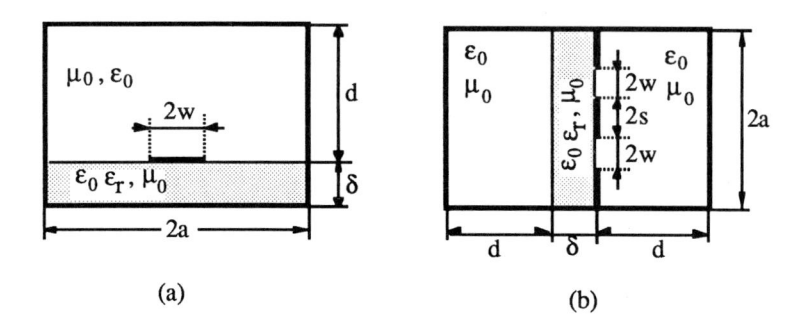

Fig. III.3 (a) Cross-section of the microstrip line and (b) cross-section of the coupled unilateral finline.

assigned **0**, in which case the results generated are for the even mode of the microstrip line or the odd mode of the slot line. Regarding the two data **NUMBER** and **FRE**, the first one should be assigned the number of the required frequencies and the second one should be assigned the actual frequencies. The number of Fourier terms n is assigned to **NN**. For a particular structure, the value of n can either be determined experimentally (by increasing n until no significant change occurs in the results computed), or it may be obtained using an empirical expression like (4.27) [4]. The order of the solution is assigned to **IN**. In fact, **IN** specifies the order of the final matrix as well as the total number of basis functions (P + Q) which enters the computation (see section 3.3.2.1). The two data **MT** and **ERM** are associated with the process of finding the roots of the determinant of the final matrix. In order to compute these roots, the program initially divides up the difference of the highest and the lowest values of the dielectric constants by an arbitrary integer **MT-1**, thereby generating **MT** fictitious dielectric constants between the two limits. The determinant values corresponding to these fictitious dielectric constants are then computed. Since at the propagation constants of the modes, the determinant vanishes, by examining the sign of the computed values of the determinant associated with the fictitious dielectric constants, the approximate places and the number of modes (roots) between the limits specified by the dielectric constants can be detected. The actual roots are subsequently computed using the bisection method. The lowest value of the dielectric constant entering the computation is **ERM** $+10^{-6}$ and the highest is **ERK** -10^{-6} where **ERK** is the largest of **ER2** and **ER1**. The value of **ERM** is assigned by the user and can be

APPENDIX III COMPUTER PROGRAMS

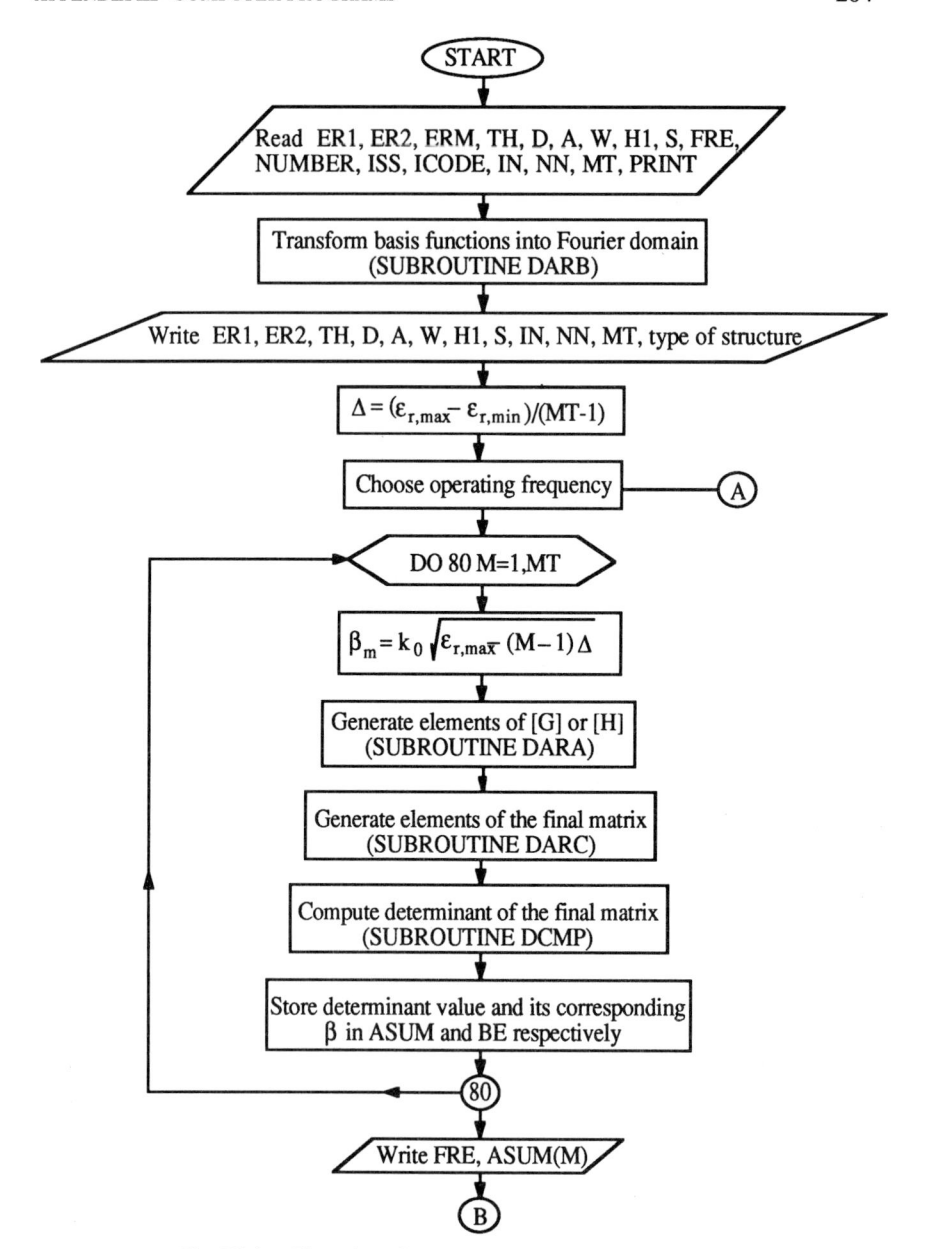

```
                    START
                      │
                      ▼
   ┌─────────────────────────────────────────┐
  / Read ER1, ER2, ERM, TH, D, A, W, H1, S, FRE, /
 /  NUMBER, ISS, ICODE, IN, NN, MT, PRINT       /
 └──────────────────────────────────────────────┘
                      │
                      ▼
      ┌──────────────────────────────────────┐
      │ Transform basis functions into Fourier domain │
      │        (SUBROUTINE DARB)             │
      └──────────────────────────────────────┘
                      │
                      ▼
  / Write ER1, ER2, TH, D, A, W, H1, S, IN, NN, MT, type of structure /
                      │
                      ▼
         ┌─────────────────────────────┐
         │ Δ = (ε_{r,max} − ε_{r,min})/(MT-1) │
         └─────────────────────────────┘
                      │
                      ▼
         ┌────────────────────────┐
         │ Choose operating frequency │──────(A)
         └────────────────────────┘
                      │
                      ▼
             ◇ DO 80 M=1,MT ◇◄──────┐
                      │             │
                      ▼             │
         ┌──────────────────────────┐│
         │ β_m = k_0 √(ε_{r,max} − (M−1)Δ) ││
         └──────────────────────────┘│
                      │             │
                      ▼             │
         ┌──────────────────────────┐│
         │ Generate elements of [G] or [H] ││
         │      (SUBROUTINE DARA)   ││
         └──────────────────────────┘│
                      │             │
                      ▼             │
         ┌──────────────────────────┐│
         │ Generate elements of the final matrix ││
         │      (SUBROUTINE DARC)   ││
         └──────────────────────────┘│
                      │             │
                      ▼             │
         ┌──────────────────────────┐│
         │ Compute determinant of the final matrix ││
         │      (SUBROUTINE DCMP)   ││
         └──────────────────────────┘│
                      │             │
                      ▼             │
         ┌──────────────────────────┐│
         │ Store determinant value and its corresponding ││
         │    β in ASUM and BE respectively    ││
         └──────────────────────────┘│
                      │             │
                      ▼             │
                    (80)────────────┘
                      │
                      ▼
           / Write FRE, ASUM(M) /
                      │
                      ▼
                     (B)
```

Fig. III.4.a Flow-chart for computing propagation constant by the spectral domain method. The rest of the chart is shown in the following figure.

APPENDIX III COMPUTER PROGRAMS 205

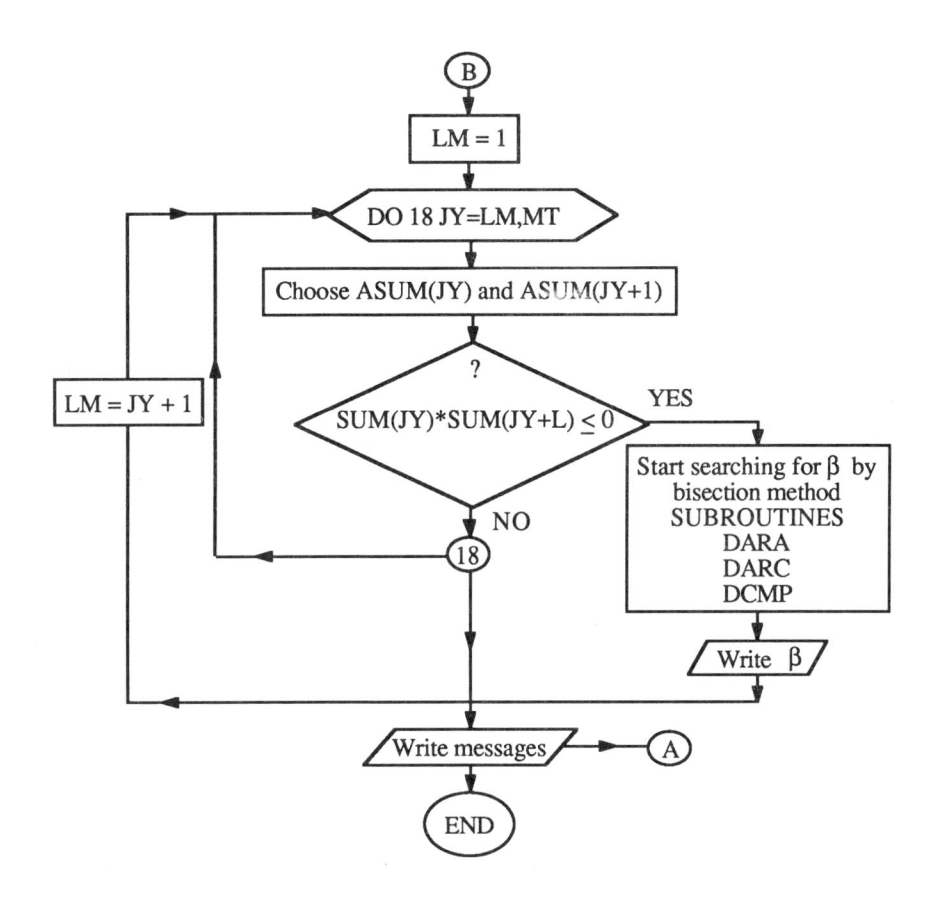

Fig. III.4.b The rest of the flow-chart for computing propagation constant by the spectral domain method.

different from the smallest dielectric constant involved in the structure. When seeking propagation constants of modes near their cut-off frequencies, **ERM** should be assigned a very small value. As regards **MT**, it should not be assigned a large value unnecessarily. A large value for **MT** could slow down the process of computing the roots. There are, however, cases where two or more roots fall within the limits specified by the dielectric constants. In such cases assigning a small value to **MT** is not recommended, since the program might miss some of the solutions. The one remaining data, **DETER**, is to disable or enable the printing process of the values of

APPENDIX III COMPUTER PROGRAMS

the determinant corresponding to the **MT** fictitious dielectric constants. These values are useful in order to separate certain spurious solutions from physical ones. Some spurious solutions can be characterised by the fact that they are poles of the determinant rather than its zeros.

After reading data, the **MAIN PROGRAM** is then engaged in the following processes.

1. It calls **SUBROUTINE DARB** to perform the Fourier transforms of the basis functions.
2. It calculates a fictitious dielectric constant and its associated propagation constant.
3. It calls **SUBROUTINE DARA** to generate the spectral domain Green's function corresponding to the above phase constant.
4. It calls **SUBROUTINE DARC** to compute the elements of the final matrix.
5. It calls **SUBROUTINE DCMP** to compute the determinant of the final matrix.

Steps 2 to 5 are repeated **MT** times and the necessary results are stored. This is clearly indicated in the flow-chart, starting at **DO 80 M=1,MT**. Following the above processing, the **MAIN PROGRAM** starts comparing the subsequent values of the determinant in order to detect a sign change. This process starts at **DO 18 JY=LM,MT** (see the flow-chart). On the detection of a sign change, indicating that there is a root between the last two values of the determinant, the program goes through the method of bisection in order to find the actual root. The accuracy in the computing of the roots is set at 0.01% in the program. This can, however, be changed to a different computing precision by replacing **1.E-4** in the following statements of the program:

    IF(EST - 1.0E-4) 22,22,23

    IF(EST - 1.0E-4) 22,22,24

After computing a root (a propagation constant), its value is printed and the program starts searching for the next root. This process is repeated until all the roots are

APPENDIX III COMPUTER PROGRAMS

computed.

As an example, a set of data together with its associated computer output for the program **MRST1** is presented in section III.6. In Fig. III.5, the values of the determinant shown in the example are plotted against the values of fictitious dielectric constants. From this figure, it is clear that one of the roots associated with the 17 GHz curve is a pole and hence, spurious.

SUBROUTINE DARB (IN,CZN,CXN)

This subroutine computes the Fourier transforms of the basis functions. The basis functions included in the program are Legendre polynomials and trigonometric functions. The former are used for approximating singular field/current distributions and the latter for approximating regular field/current distributions. For further information on this matter, the reader may refer to section 4.1.

In the subroutine, **IN** represents the total number of basis functions. Each singular or regular field/current distribution is expanded in terms of **IN**/2 basis functions (P=Q=**IN**/2). All the data (except **IN**) and the other required parameters are passed to this subroutine through three common blocks (see the program listing). At the output, arrays **CZN** and **CXN** contain the Fourier spectra of the basis functions. Note that in this subroutine the arrays' dimensions are fixed, ie. **CZN(7,200)**, **CXN(7,200)**. When the number of Fourier terms $n>200$, these dimensions must be modified.

SUBROUTINE DARA (BET,FREQ,GG11,GG12,GG21,GG22)

This subroutine computes the spectral domain dyadic Green's function of the three-layer structure when the third layer is to be air, Fig. III.6. According to the solution requirement, which can be set by **ISS**, the program either computes the elements of $[G(\alpha_n,\beta)]$ given by Eqs. (6.25) or it calculates the elements of $[H(\alpha_n,\beta)]$. In the former case **ISS** is **6** and in the latter case **ISS** is **5**. In either case, the subroutine can generate the even form (when **ICODE** is **1**) or the odd form (when **ICODE** is **-1**) of the Green's function. On exit, arrays **GG11, GG12, GG21** and **GG22** contain the elements of $[G(\alpha_n,\beta)]$ or $[H(\alpha_n,\beta)]$. We recall that $[G(\alpha_n,\beta)]$ is used when current distributions $J_x(x)$ and $J_z(x)$ are expanded in terms of basis functions whereas $[H(\alpha_n,\beta)]$ is employed in the solution when field

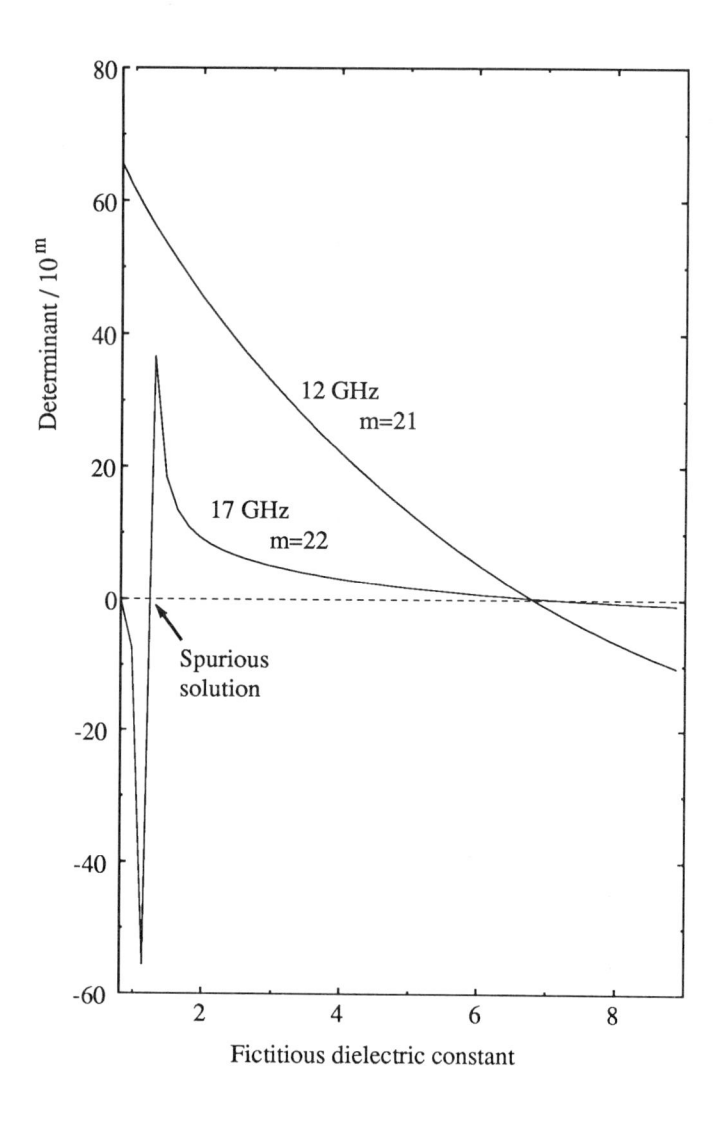

Fig. III.5 Variation of the determinant with fictitious dielectric constant for the example presented in section III.6.

APPENDIX III COMPUTER PROGRAMS

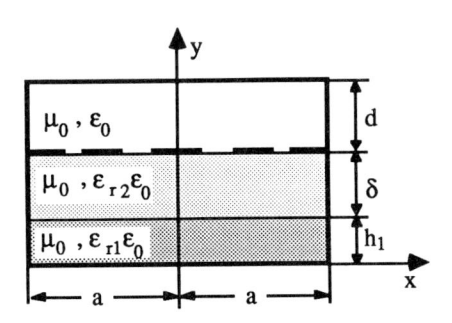

Fig. III.6 Generic cross-section of transmission lines whose Green's functions can be computed using SUBROUTINE DARA. The cross-section is symmetrical with respect to the y-axix.

distributions $E_x(x)$ and $E_z(x)$ are approximated by sets of basis functions. For further information on the applications of the mentioned matrices, the reader may refer to section 4.4. All data except **BET** and **FREQ**, together with the parameters π (**PI**), ε_0 (**EØ**) and μ_0 (**AMØ**) are passed to the subroutine through common blocks. Note that if the computations require $n>200$ (n is the number of Fourier terms which is shown in the program by **NN**), the dimensions of **GG11**, **GG12** etc. must be adjusted accordingly.

SUBROUTINE DARC (CZN,CXN,GG11,GG12,GG21,GG22,SZ,IN,NN)

This subroutine computes the elements of the final matrix and return the results in array **SZ**. Inputs to this subroutine are the elements of the spectral domain Green's functions **GG11**, **GG12** etc., the Fourier transforms of basis functions **CZN**, **CXN**, the order of the matrix **IN** and the number of Fourier terms **NN**. The value of **IN** is dictated by the number of basis functions involved in the computation and is set by the user. Note that for **NN>200**, the dimensions of the arrays used in the subroutine must be modified.

SUBROUTINE DCMP (IN,Y,IVT,DET)

This subroutine, which is based on the Gauss elimination technique, computes the

APPENDIX III COMPUTER PROGRAMS 210

value of the determinant of the final matrix. On exit, **DET** contains this value. Inputs to the subroutine are the order of the matrix **IN** and the elements of the matrix **Y**. With the specified dimensions, the subroutine can be used for matrices as large as 20 × 20. For the calculation of the determinants of matrices larger than 20 × 20, the dimensions of the declared arrays in the subroutine must be adjusted. Note that array **IVT** is neither an input nor an output array, but it is working space which must be declared in order that the subroutine works satisfactorily.

III.2 SUBROUTINE MGNTW (BET,FREQ,GG11,GG12,GG21,GG22)

Subroutine **MGNTW** (see section III.7 for the listing) generates the spectral domain Green's function of three-layer planar transmission lines with a magnetic wall at the bottom of the enclosure, Fig. III.2.a. Note that the program generates the elements of $[H(\alpha_n,\beta)]$ only. It can, however, be easily modified to compute other forms of Green's functions; eg: $[G(\alpha_n,\beta)]$. The structure of the program is essentially the same as **DARA** explained in the previous section, and so are the structures of the inputs and the outputs. Particularly note that on exit, arrays **GG11**, **GG12** etc., contain the elements of $[H(\alpha_n,\beta)]$. The subroutine can be linked to **DARB**, **DARC** and **DCMP** in order to compute dispersion characteristics of bilateral finlines etc.

III.3 SUBROUTINE ANPDL (BET,FREQ,GG11,GG12,GG21,GG22)

Subroutine **ANPDL** (see section III.8 for the listing) generates the spectral domain Green's function of planar transmission lines with rotational symmetry. In Fig. III.2.b, the generic cross-section of these lines is shown. Each half of the structure which is separated by the $y=0$ plane, consists of three dielectric layers. The program assumes that the top and the bottom layers are air (ε_0, μ_0). Since the program is originally developed for the analysis of antipodal finlines, (see section 6.2.3), it generates the elements of Green's function $[H(\alpha_n,\beta)]$. However, if required, it is easy to modify the program in order that it computes other forms of Green's functions; eg: $[G(\alpha_n,\beta)]$. The input and the output parameters of the program are similar to those explained for **DARA** in section III.1. Particularly note that on exit, arrays **GG11**, **GG12**, etc. contain the elements of $[H(\alpha_n,\beta)]$.

APPENDIX III COMPUTER PROGRAMS

III.4 SUBROUTINE MSCL2 (BET,FREQ,GG11,GG12,GG13,GG14, GG22,GG23,GG31,GG32,GG41)

Subroutine **MSCL2** (see section III.9 for the listing) can be used to compute the elements of the spectral domain Green's function $(I_{11}(\alpha_n,\beta), I_{12}(\alpha_n,\beta)$ etc.) introduced in Eq. (6.44). These elements are given by expressions (6.45). A wide range of transmission lines of the generalised cross-section shown in Fig. III.2.c can be analysed by the spectral domain technique using this subroutine. The input and output parameters of the subroutine are similar to those explained for **DARA** in section III.1. Minor differences between inputs of **MSCL2** and **DARA** are as follows: **H1** in **DARA** is **H** in **MSCL2** and no **ER3** exists in **DARA**. With reference to Fig. III.2.c, **ER3** and **H** are computer equivalents of ε_{r3} and h respectively. On exit, arrays **GG11, GG12, GG13, GG14, GG22, GG23, GG31, GG32** and **GG41** contain elements $I_{11}(\alpha_n,\beta), I_{12}(\alpha_n,\beta), I_{13}(\alpha_n,\beta), I_{14}(\alpha_n,\beta), I_{22}(\alpha_n,\beta), I_{23}(\alpha_n,\beta), I_{31}(\alpha_n,\beta), I_{32}(\alpha_n,\beta)$ and $I_{41}(\alpha_n,\beta)$ respectively. Using the relations in (6.45) between the elements, all the sixteen elements of the spectral domain matrix introduced in (6.44) can be computed.

III.5 Listing of MRST1

```
C**************************************************
C                                                  *
C                                                  *
C                                                  *
C             PROGRAM NAME:    MRST1               *
C             ----------------------------         *
C                                                  *
C                                                  *
C             PROGRAMMING LANGUAGE    FORTRAN      *
C             PRECISION              DOUBLE        *
C             CORE REQUIREMENT    <  120 KB        *
C                                                  *
C                                                  *
C      THIS COMPUTER PROGRAM COMPUTES DISPERSION   *
C   CHARACTERISTICS OF SOME USEFUL TRANSMISSION    *
C   LINES INCLUDING                                *
C                                                  *
C   1. SHIELDED MICROSTRIP LINES IN CONVENTIONAL AND *
C      INVERTED FORMS,                             *
C   2. SHIELDED COUPLED MICROSTRIP LINES IN        *
C      CONVENTIONAL AND INVERTED FORMS (BOTH THE EVEN AND *
C      THE ODD MODES),                             *
C   3. SHIELDED SLOT LINES AND UNILATERAL FINLINES, *
C   4. SHIELDED COUPLED SLOT LINES, COUPLED FINLINES AND *
C      COPLANAR WAVEGUIDES (BOTH THE EVEN MODE AND THE *
C      ODD MODES).                                 *
C                                                  *
C      THE METHOD OF SOLUTION IS THE SPECTRAL DOMAIN *
C      TECHNIQUE BASED ON THE TRANSFER MATRIX APPROACH. *
C                                                  *
C      WITH REFERENCE TO FIG.III.1, THE PROGRAM REQUIRES *
C      THE FOLLOWING DATA:                         *
C                                                  *
C   ER1      RELATIVE PERMITTIVITY OF SUBSTRATE 1  *
C   ER2      RELATIVE PERMITTIVITY OF SUBSTRATE 2  *
C   ERM      MINIMUM RELATIVE PERMITTIVITY         *
C   TH       THICKNESS OF SUBSTRATE 2              *
C   D        HEIGHT OF THE ENCLOSURE FROM TOP OF   *
C            SUBSTRATE 2                           *
C   A        HALF-WIDTH OF THE ENCLOSURE           *
C   W        HALF-WIDTH OF MICROSTRIP(S) OR SLOT(S) (SEE *
C            THE PROGRAM DESCRIPTION.)             *
C   H1       THICKNESS OF SUBSTRATE 1              *
C   S        HALF-WIDTH OF SLOT(S) OR MICROSTRIP(S) (SEE *
C            THE PROGRAM DESCRIPTION.)             *
C   ICODE    -1,0,1(ODD MODE,SINGLE LINE,EVEN MODE) *
C   ISS      5 (FOR SLOT LINES),6 (FOR STRIP LINES) *
C   NUMBER   NUMBER OF REQUIRED FREQUENCIES < 10   *
C   IN       SOLUTION ORDER ('IN' TAKES EVEN VALUES ONLY; *
```

APPENDIX III COMPUTER PROGRAMS

```
C               CHOOSE 'IN' FROM 4 TO 14 FOR COUPLED LINES  *
C               AND FROM 4 TO 8 FOR SINGLE LINES.)          *
C     NN        NUMBER OF FOURIER TERMS<200                 *
C     MT        RESOLUTION PARAMETER<50                     *
C     PRINT     'Y' OR 'N' (PRINT OR NOT PRINT THE DETERMINANT*
C               OF THE FINAL MATRIX.)                       *
C     FRE       REQUIRED FREQUENCIES IN HZ                  *
C                                                           *
C REMARKS:                                                  *
C                                                           *
C  *  'IN' AND 'NN' HAVE IMPORTANT ROLES IN BOTH ACCURACY*
C     AND CPU TIMES. USUALLY IN=6 AND NN=100 CAN BE USED    *
C     WHEN A/W<10. AS THIS RATIO OR 'IN' IS INCREASED,      *
C     'NN' SHOULD BE RAISED. NOTE THAT WHENEVER THE NO.     *
C     OF SOLUTIONS IS MORE THAN EXPECTED, INCREASE 'NN'.    *
C                                                           *
C  *  'MT' IS INTENDED TO RESOLVE SOLUTIONS WHENEVER        *
C     THEY OVERLAP.                                         *
C                                                           *
C  *  FOR THE VALUE OF ERM, SEE THE PROGRAM DESCRIPTION.    *
C                                                           *
C  *  BY APPROPRIATELY LOCATING SIDE WALLS , TOP AND/OR     *
C     BOTTOM WALL, THE PROGRAM CAN BE USED FOR OPEN         *
C     STRUCTURES WITH SUFFICIENT ACCURACY.                  *
C                                                           *
C  *  NEVER ASSIGN ZERO TO 'TH'. WHENEVER SINGLE            *
C     SUBSTRATE IS INVOLVED, ASSIGN ZERO TO 'H1'.           *
C                                                           *
C  *  ALL DIMENSIONS IN MILLIMETER(MM).                     *
C***********************************************************
C*************************************************
C THE ORDER AND FORMATS OF DATA ARE AS FOLLOWS:*
C                                              *
C                                              *
C***********************************************************
C             DATA AND FORMATS                             *
C***********************************************************
C   DATA      *  DATA NO.  *  COLUMN  *  FORMAT      *
C***********************************************************
C   ER1       *     1      *   1-20   *  F20.16      *
C***********************************************************
C   ER2       *     2      *   1-20   *  F20.16      *
C***********************************************************
C   ERM       *     3      *   1-20   *  F20.16      *
C***********************************************************
C   TH        *     4      *   1-20   *  F20.16      *
C***********************************************************
C   D         *     5      *   1-20   *  F20.16      *
C***********************************************************
C   A         *     6      *   1-20   *  F20.16      *
C***********************************************************
C   W         *     7      *   1-20   *  F20.16      *
```

APPENDIX III COMPUTER PROGRAMS

```
C*****************************************************************
C     H1         *    8     *   1-20   *    F20.16       *
C*****************************************************************
C     S          *    9     *   1-20   *    F20.16       *
C*****************************************************************
C     ICODE      *   10     *   1-4    *    I4           *
C*****************************************************************
C     ISS        *   10     *   5-8    *    I4           *
C*****************************************************************
C     NUMBER     *   10     *   9-12   *    I4           *
C*****************************************************************
C     IN         *   11     *   1-4    *    I4           *
C*****************************************************************
C     NN         *   11     *   5-8    *    I4           *
C*****************************************************************
C     MT         *   11     *   9-12   *    I4           *
C*****************************************************************
C     PRINT      *   12     *   1      *    A1           *
C*****************************************************************
C     FRE        *   13     *   1-20   *    F20.16       *
C*****************************************************************
      IMPLICIT REAL*8(A-H,O-Z)
      DIMENSION BE(50),XNG(50),ASUM(50),CZN(7,200),CXN(7,200)
      DIMENSION FRE(10),GG11(200),GG12(200),GG21(200),GG22(200)
      DIMENSION SZ(20,20),Y(20,20),EA(50)
      INTEGER IVT(20)
      CHARACTER PRINT,YES
      COMMON/XXX/E0,ER1,ER2,AM0,H1,TH,D,A,W,S,NN
      COMMON/XX/ICODE,ISS
      COMMON PI
      DATA YES/'Y'/
      OPEN(3,file='IN')
      OPEN(4,file='OUT')
      READ(3,1) ER1,ER2,ERM,TH,D,A,W,H1,S
      READ(3,40) ICODE,ISS,NUMBER,IN,NN,MT,PRINT
   40 FORMAT(3I4/3I4/A1)
    1 FORMAT(F20.16)
      PI=3.14159265358979D0
      AM0=12.56D-10
      E0=8.854D-15
      EEE=ER1
      SME=DSQRT(AM0*E0)
C*****************************************************************
C                                                                 *
C     READ FREQUENCIES                                            *
C     ----------------------                                      *
C
      READ(3,6) (FRE(K),K=1,NUMBER)
    6 FORMAT(F20.16)
C*****************************************************************
      CALL DARB(IN,CZN,CXN)
      IF(H1-1.D-5) 150,151,151
```

APPENDIX III COMPUTER PROGRAMS

```
    150  ER1=ER2
         GO TO 152
    151  ER1=EEE
    152  WRITE(4,200)ER1,ER2,H1,TH,D,A,W,S
         WRITE(4,260)NN,IN,MT
    200  FORMAT(' ER1=',F6.3/' ER2=',F6.3/
        *' H1=',F10.7,' MM'/' TH=',F10.7,' MM'/
        *' D=',F10.7,' MM'/' A=',F10.7,' MM'/
        *' W=',F10.7,' MM'/' S=',F10.7,' MM'/)
    260  FORMAT(' NN=',I4,2X,' IN=',I4,2X,' MT=',I4)
C    ---------------------------------------------
         IF(ICODE.EQ.-1.AND.ISS.EQ.6) WRITE(4,44)
         IF(ICODE.EQ.0.AND.ISS.EQ.6) WRITE(4,45)
         IF(ICODE.EQ.1.AND.ISS.EQ.6) WRITE(4,47)
         IF(ICODE.EQ.-1.AND.ISS.EQ.5) WRITE(4,42)
         IF(ICODE.EQ.0.AND.ISS.EQ.5) WRITE(4,43)
         IF(ICODE.EQ.1.AND.ISS.EQ.5) WRITE(4,48)
     44  FORMAT(//10X,'COUPLED STRIPS (ODD MODE)'/
        *         10X,'_____'//)
     45  FORMAT(//10X,'SINGLE STRIP'/
        *         10X,'_____'//)
     47  FORMAT(//10X,'COUPLED STRIPS (EVEN MODE)'/
        *         10X,'_____'//)
     42  FORMAT(//10X,'COUPLED SLOTS (ODD MODE)'/
        *         10X,'_____'//)
     43  FORMAT(//10X,'SINGLE SLOT'/
        *         10X,'_____'//)
     48  FORMAT(//10X,'COUPLED SLOTS (EVEN MODE)'/
        *         10X,'_____'//)
C    ---------------------------------------------
         TM=MT-1
         IF(ER2-ER1) 30,30,32
     32  ERK=ER2-1.D-6
         GO TO 34
     30  ERK=ER1-1.D-6
     34  DIV=(ERK-ERM-1.D-6)/TM
C    --------------------
     46  DO 8 JJ=1,NUMBER
         FREQ=FRE(JJ)
         AK0=2.D0*PI*FREQ*SME
C    --------------------
         DO 80 M=1,MT
         ERV=ERK-DFLOAT(M-1)*DIV
         EA(M)=ERV
         WRITE(6,600)ERV
    600  FORMAT(' ERV=',F15.7)
         BE(M)=AK0*DSQRT(ERV)
         BET=BE(M)
         CALL DARA(BET,FREQ,GG11,GG12,GG21,GG22)
         CALL DARC(CZN,CXN,GG11,GG12,GG21,GG22,SZ,IN,NN)
         DO 5 JI=1,IN
         DO 5 KI=1,IN
```

APPENDIX III COMPUTER PROGRAMS

```
    5 Y(JI,KI)=SZ(JI,KI)
      CALL DCMP(IN,Y,IVT,DET)
      SUM=DET
      ASUM(M)=SUM
      XNG(M)=AK0/BET
   80 CONTINUE
      WRITE(4,15)FRE(JJ)
   15 FORMAT(' FREQ=',D15.7,' HZ')
C************************************************************
C   THE FOLLOWING STATEMENT WRITES DETERMINANT VALUES.     *
      IF(PRINT.EQ.YES)WRITE(4,99)(EA(MR),ASUM(MR),MR=1,MT)
   99 FORMAT(/2X,'FICTITIOUS DIELECTRIC CONSTANT',
     *10X,'DETERMINANT'/ (10X,F15.3,14X,D15.7))
C************************************************************
      LM=1
   70 DO 18 JY=LM,MT
      IF(JY+1-MT)83,83,86
   83 SUM1=ASUM(JY)
      BET1=BE(JY)
      SUM2=ASUM(JY+1)
      BET2=BE(JY+1)
      X=SUM2/SUM1
      IF(X) 19,19,18
   18 CONTINUE
   19 BET=(BET1+BET2)/2.D0
      CALL DARA(BET,FREQ,GG11,GG12,GG21,GG22)
      CALL DARC(CZN,CXN,GG11,GG12,GG21,GG22,SZ,IN,NN)
      DO 7 JB=1,IN
      DO 7 KB=1,IN
    7 Y(JB,KB)=SZ(JB,KB)
      CALL DCMP(IN,Y,IVT,DET)
      BSUM=DET
      IF(BSUM/SUM1) 20,21,21
   20 EST=DABS(BET-BET1)/BET
      IF(EST-1.0E-4) 22,22,23
   21 EST=DABS(BET-BET2)/BET
      IF(EST-1.0E-4) 22,22,24
   23 BET2=BET
      GO TO 19
   24 BET1=BET
      SUM1=BSUM
      GO TO 19
   22 XNT=AK0/BET
      VXNT=1.D0/XNT
      EEFF=VXNT**2
      WRITE(4,25) BET,VXNT,XNT,EEFF
   25 FORMAT(/' BETA=',F20.14,' RAD./MM'/
     *        ' NORMALIZED BETA=',F10.5/
     *        ' NORMALIZED WAVE-LENGTH=',F10.5/
     *        ' E EFFECTIVE=',F10.5/)
C     ---------------------------------------------------
      LM=JY+1
```

APPENDIX III COMPUTER PROGRAMS

```
         GO TO 70
   86    WRITE(4,87)
   87    FORMAT(/ 'CHECK THE DETERMINANT FOR ANY SPURIOUS MODE.'//)
    8    CONTINUE
         STOP
         END
C*************************************************************
C                                                             *
C  THE FOLLOWING SUBROUTINE COMPUTES THE SPECTRAL DOMAIN      *
C  GREEN'S FUNCTION OF THREE-LAYER PLANAR TRANSMISSION        *
C  LINES OF GENERAL CROSS-SECTION SHOWN IN FIG.III.6 OF       *
C  THIS APPENDIX.
C                                                             *
         SUBROUTINE DARA(BET,FREQ,GG11,GG12,GG21,GG22)
C                                                             *
C*************************************************************
         IMPLICIT REAL*8(A-H,O-Z)
         DIMENSION GG11(200),GG12(200),GG21(200),GG22(200)
         COMMON PI
         COMMON/XXX/E0,ER1,ER2,AM0,H1,TH,D,A,W,S,NN
         COMMON/XX/ICODE,ISS
         ER02=ER2*E0
         OMEGA=2.*PI*FREQ
         AK0=OMEGA*DSQRT(AM0*E0)
         AK1=AK0*DSQRT(ER1)
         AK2=AK0*DSQRT(ER2)
         SK0=AK0**2
         SK1=AK1**2
         SK2=AK2**2
         SBET=BET**2
         AKB0=SK0-SBET
         AKB1=SK1-SBET
         AKB2=SK2-SBET
         AKOB=AKB0/BET
         SKK=SK1-SK2
         ICCD=ICODE
         DO 100 N=1,NN
C*************************************************************
C                                                             *
C  NOTE HOW THE SPECTRAL DOMAIN PARAMETER 'ALPN'              *
C  IS ADJUSTED ACCORDING TO THE PROBLEM REQUIREMENT (SEE      *
C  EQS.(6.26)).                                               *
C                                                             *
         IF(ISS.EQ.5.AND.ICODE.EQ.0)ICCD=-1
         IF(ICCD) 1,2,2
    2    AN=N
         ALPN=(AN-0.5D0)*PI/A
         GO TO 3
    1    AN=N-1
C*************************************************************
         ALPN=AN*PI/A
    3    SALPN=ALPN**2
```

APPENDIX III COMPUTER PROGRAMS

```
C     --------------
      ABA=SALPN+SBET-SK2
      GAMA2=DSQRT(DABS(ABA))
      GH2=GAMA2*TH
      IF(ABA) 50,50,51
   50 TA2=DSIN(GH2)/DCOS(GH2)
      GO TO 52
   51 TA2=DTANH(GH2)
   52 ACA=SALPN+SBET-SK1
      ADA=SALPN+SBET-SK0
      GAMA1=DSQRT(DABS(ACA))
      GAMA0=DSQRT(DABS(ADA))
      GH1=GAMA1*H1
      GD=GAMA0*D
C     --------------
      GAA=GAMA1/GAMA2
      EG12=GAA*ER1/ER2
      ABOG=ALPN*BET/(OMEGA*GAMA2)
      ABOGE=ABOG*ER02
      ABOGM=ABOG/AM0
      OBG2=OMEGA*BET*GAMA2
      OBG2E=OBG2*ER02
      OBG2M=OBG2*AM0
      OBG0E=OMEGA*E0*GAMA0/BET
      OBG0M=OMEGA*AM0*GAMA0/BET
      AKG2=SK2*GAMA2**2
      AMUL=SK2*AKB1-SALPN*SK1
C     ------------------------
      IF(ACA) 60,60,61
   61 TA1=DTANH(GH1)
      D11=(AKB1+GAA*AKB2*TA1*TA2)/BET
      F11=-(AMUL*TA2-AKG2*GAA*TA1)/OBG2E
      H11=-ALPN*(1.D0+GAA*TA2*TA1)
      GO TO 62
   60 TA1=DSIN(GH1)/DCOS(GH1)
      D11=(AKB1-GAA*AKB2*TA1*TA2)/BET
      F11=-(AMUL*TA2+AKG2*GAA*TA1)/OBG2E
      H11=-ALPN*(1.D0-GAA*TA2*TA1)
   62 A11=(AKB1*TA1+EG12*AKB2*TA2)/BET
      B11=ABOGE*SKK*TA2/BET
      C11=-ABOGM*SKK*TA2*TA1/BET
      AL1=-ALPN*(TA1+EG12*TA2)
      G11=(AMUL*TA2*TA1-AKG2*EG12)/OBG2M
C     -------------------------------------
      IF(ADA) 63,63,64
   64 CTAD=1.D0/DTANH(GD)
      H0=-ALPN*CTAD
      B0=AKOB*CTAD
      GO TO 65
   63 CTAD=DCOS(GD)/DSIN(GD)
      H0=ALPN*CTAD
      B0=-AKOB*CTAD
```

APPENDIX III COMPUTER PROGRAMS

```
   65 A0=AKOB
      C0=-ALPN
      D0=-OBG0M
      AL0=OBG0E*CTAD
C     -------------------
      ADBC=A11*D11+B11*C11
      AHBG=A11*H11+B11*G11
      BLAF=B11*AL1-A11*F11
      CHGD=C11*H11-G11*D11
      CFLD=C11*F11+AL1*D11
      GFLH=G11*F11+AL1*H11
C     -------------------
      CH0=C0*H0
      DL0=D0*AL0
      CB0=C0*B0
      BL0=AL0*B0
      AH0=A0*H0
      AD0=A0*D0
      AB0=A0*B0
C     -------------------
      DET=ADBC*(CH0+DL0)-AHBG*CB0+BLAF*BL0+CHGD*AD0
     5-CFLD*AH0+GFLH*AB0
      AM1=(AHBG*D0+BLAF*H0)/DET
      AN1=-(BLAF*B0+ADBC*D0)/DET
      AM2=(GFLH*A0+BLAF*AL0-AHBG*C0)/DET
      AN2=(ADBC*C0-CFLD*A0)/DET
      DD11=-AKOB*AM1
      DD12=AKOB*AN1
      DD21=ALPN*AM1+OBG0M*AM2
      DD22=-ALPN*AN1-OBG0M*AN2
      DETG=DD11*DD22-DD12*DD21
C     -----------------------
      IF(ISS.EQ.6)GO TO 98
      GG11(N)=DD22/DETG
      GG12(N)=-DD12/DETG
      GG21(N)=-DD21/DETG
      GG22(N)=DD11/DETG
      IF(ICCD.GT.0)GOTO 100
      IF(N.EQ.1)GG12(N)=GG12(N)/2.D0
      GO TO 100
C     -----------------------
   98 GG11(N)=DD11
      GG12(N)=DD12
      GG21(N)=DD21
      GG22(N)=DD22
      IF(ICCD) 4,100,100
    4 IF(N-1)200,200,100
  200 GG21(N)=GG21(N)/2.D0
      GG11(N)=GG11(N)/2.D0
  100 CONTINUE
      RETURN
      END
```

APPENDIX III COMPUTER PROGRAMS

```
C***********************************************************
C                                                            *
C     THE FOLLOWING SUBROUTINE COMPUTES THE FOURIER          *
C     TRANSFORMS OF BASIS FUNCTIONS.                         *
C                                                            *
      SUBROUTINE DARB(IN,CZN,CXN)
C                                                            *
C***********************************************************
      IMPLICIT REAL*8(A-H,O-Z)
      DIMENSION CZN(7,200),CXN(7,200)
      COMMON PI
      COMMON/XXX/E0,ER1,ER2,AM0,H1,TH,D,A,W,S,NN
      COMMON/XX/ICODE,ISS
      IN2=IN/2
      ICCD=ICODE
      DO 2 N=1,NN
      IF(ISS.EQ.5.AND.ICODE.EQ.0)ICCD=-1
      IF(ICCD) 8,9,9
C     --------------------
    9 AN=N
      ALPN=(AN-0.5D0)*PI/A
      GO TO 10
    8 AN=N-1
      ALPN=AN*PI/A
   10 PN=ALPN*W
      QN=ALPN*(S+W)
      SINP=DSIN(PN)
      COSP=DCOS(PN)
      SINQ=DSIN(QN)
      COSQ=DCOS(QN)
      IF(ICCD) 16,4,4
   16 IF(N-1) 3,3,4
    3 AI0=2.D0
      AI1=0.D0
      AI2=2.D0/3.D0
      AI3=0.D0
      AI4=2.D0/5.D0
      AI5=0.D0
      AI6=2.D0/7.D0
      IF(ISS.EQ.5.AND.ICODE.EQ.0)GOTO 30
      IF(ISS.EQ.5)GOTO 18
      GO TO 5
    4 SIDP=2.D0*SINP/PN
      CODP=2.D0*COSP/PN
      AI0=SIDP
      AI1=-CODP+AI0/PN
      AI2=SIDP-2.D0*AI1/PN
      AI3=-CODP+3.D0*AI2/PN
      AI4=SIDP-4.D0*AI3/PN
      AI5=-CODP+5.D0*AI4/PN
      AI6=SIDP-6.D0*AI5/PN
C     --------------------
```

APPENDIX III COMPUTER PROGRAMS

```
        IF(ISS.EQ.6)GOTO 40
        IF(ICODE) 18,30,5
  40    IF(ICODE) 5,30,18
  18    CZN(1,N)=AI0*COSQ
        CZN(2,N)=-AI1*SINQ
        CZN(3,N)=0.5D0*(3.D0*AI2-AI0)*COSQ
        CZN(4,N)=-0.5D0*(5.D0*AI3-3.D0*AI1)*SINQ
        CZN(5,N)=0.125D0*(35.D0*AI4-30.D0*AI2+3.D0*AI0)*COSQ
        CZN(6,N)=-0.125D0*(63.D0*AI5-70.D0*AI3+15.D0*AI1)*SINQ
        CZN(7,N)=0.0625D0*(231.D0*AI6-315.D0*AI4+105.D0*AI2
      9 -5.D0*AI0)*COSQ
        GO TO 20
C       ------------
  30    CZN(1,N)=AI0
        CZN(2,N)=0.5D0*(3.D0*AI2-AI0)
        CZN(3,N)=0.125D0*(35.D0*AI4-30.D0*AI2+3.D0*AI0)
        CZN(4,N)=0.0625D0*(231.D0*AI6-315.D0*AI4+105.D0*AI2-5.D0*AI0)
        GO TO 20
C       ------------------------------------------------------------
  5     CZN(1,N)=AI0*SINQ
        CZN(2,N)=AI1*COSQ
        CZN(3,N)=0.5D0*(3.D0*AI2-AI0)*SINQ
        CZN(4,N)=0.5D0*(5.D0*AI3-3.D0*AI1)*COSQ
        CZN(5,N)=0.125D0*(35.D0*AI4-30.D0*AI2+3.D0*AI0)*SINQ
        CZN(6,N)=0.125D0*(63.D0*AI5-70.D0*AI3+15.D0*AI1)*COSQ
        CZN(7,N)=0.0625D0*(231.D0*AI6-315.D0*AI4+105.D0*AI2
      8 -5.D0*AI0)*SINQ
  20    DO 2 M=1,IN2
        AM=M
        APN=AM*PI+PN
        AMN=AM*PI-PN
        SINMM=DSIN(AMN)
        SINMP=DSIN(APN)
        COSMM=DCOS(AMN)
        COSMP=DCOS(APN)
        TM=AM*PI-2.D0*PN
        TP=AM*PI+2.D0*PN
        SI1=(SINP+SINMP)/TP+(SINP-SINMM)/TM
        SI2=(COSP-COSMP)/TP+(COSP-COSMM)/TM
        IF(ISS.EQ.6)GOTO 42
        IF(ICODE) 12,32,11
  42    IF(ICODE) 11,32,12
  12    CXN(M,N)=SI1*COSQ+SI2*SINQ
        GO TO 2
  32    CXN(M,N)=SINMM/AMN-SINMP/APN
        GO TO 2
  11    CXN(M,N)=-SI1*SINQ+SI2*COSQ
  2     CONTINUE
 100    CONTINUE
        RETURN
        END
```

APPENDIX III   COMPUTER PROGRAMS   222

```
C**************************************************************
C                                                              *
C   THE FOLLOWING SUBROUTINE COMPUTES THE ELEMENTS OF THE      *
C   FINAL MATRIX.                                              *
C                                                              *
      SUBROUTINE DARC(CZN,CXN,GG11,GG12,GG21,GG22,SZ,IN,NN)
C                                                              *
C**************************************************************
      IMPLICIT REAL*8(A-H,O-Z)
      DIMENSION CZN(7,200),CXN(7,200),GG11(200),GG12(200),GG21(200)
      DIMENSION GG22(200),Z(20,20),SZ(20,20)
      IN2=IN/2
      IN22=IN2+1
      DO 20 M=1,IN
      DO 20 L=1,IN
   20 SZ(M,L)=0
      DO 10 N=1,NN
      DO 53 IM=1,IN2
      DO 15 IO=1,IN2
   15 Z(IM,IO)=CZN(IM,N)*GG11(N)*CXN(IO,N)
      DO 53 IOO=IN22,IN
   53 Z(IM,IOO)=CZN(IM,N)*GG12(N)*CZN(IOO-IN2,N)
   42 DO 63 IMM=IN22,IN
   12 DO 16 IIO=1,IN2
   16 Z(IMM,IIO)=CXN(IMM-IN2,N)*GG21(N)*CXN(IIO,N)
      DO 63 IIOO=IN22,IN
   63 Z(IMM,IIOO)=CXN(IMM-IN2,N)*GG22(N)*CZN(IIOO-IN2,N)
      DO 10 J=1,IN
      DO 10 I=1,IN
      SZ(I,J)=SZ(I,J)+Z(I,J)
   10 CONTINUE
      RETURN
      END
C**********************************************
C                                              *
C THE FOLLOWING SUBROUTINE COMPUTES THE        *
C DETERMINANT OF THE FINAL MATRIX USING        *
C THE GAUSS ELIMINATION TECHNIQUE.             *
C                                              *
      SUBROUTINE DCMP(IN,Y,IVT,DET)
C                                              *
C**********************************************
      IMPLICIT REAL*8(A-H,O-Z)
      DIMENSION Y(20,20)
      INTEGER IVT(20)
      N=IN
      IVT(N)=1
      NM1=N-1
      ANRM=0.0D0
      DO 10 J=1,N
      U=0.D0
      DO 5 I=1,N
```

APPENDIX III COMPUTER PROGRAMS

```
    5   U=U+DABS(Y(I,J))
   10   IF(U.GT.ANRM)ANRM=U
        DO 27 K=1,NM1
        KP1=K+1
        M=K
        DO 15 I=KP1,N
   15   IF(DABS(Y(I,K)).GT.DABS(Y(M,K)))M=I
        IVT(K)=M
        IF(M.NE.K)IVT(N)=-IVT(N)
        U=Y(M,K)
        Y(M,K)=Y(K,K)
        Y(K,K)=U
        IF(U.EQ.0.D0)GOTO 27
        DO 20 I=KP1,N
   20   Y(I,K)=-Y(I,K)/U
        DO 26 J=KP1,N
        U=Y(M,J)
        Y(M,J)=Y(K,J)
        Y(K,J)=U
        IF(U.EQ.0.D0)GOTO 26
        DO 25 I=KP1,N
        Y(I,J)=Y(I,J)+Y(I,K)*U
   25   CONTINUE
   26   CONTINUE
   27   CONTINUE
        DET=DBLE(FLOAT(IVT(N)))
        DO 30 JI=1,N
   30   DET= DET*Y(JI,JI)
        RETURN
        END
```

APPENDIX III COMPUTER PROGRAMS   224

III.6 Sample Input and Output

A sample input for the computer program MRST1

```
1.0000
8.8750
0.8000
1.2700
11.430
6.3501
0.6350
0.0000
0.0000
    0   6   2
    8 180  50
Y
12000000000.
17000000000.
```

Output for the above sample input

```
ER1= 8.875
ER2= 8.875
H1= 0.0000000 MM
TH= 1.2700000 MM
D=11.4300000 MM
A= 6.3501000 MM
W= 0.6350000 MM
S= 0.0000000 MM

NN= 180    IN=   8    MT=   50


           SINGLE STRIP
           _____


FREQ=  0.1200000d+11  HZ

   FICTITIOUS DIELECTRIC CONSTANT           DETERMINANT
                    8.875                 -0.1038326d+23
                    8.710                 -0.9704895d+22
                    8.545                 -0.9001274d+22
                    8.381                 -0.8272055d+22
                    8.216                 -0.7516889d+22
                    8.051                 -0.6735413d+22
                    7.886                 -0.5927255d+22
                    7.721                 -0.5092030d+22
```

APPENDIX III COMPUTER PROGRAMS

7.557	-0.4229341d+22
7.392	-0.3338776d+22
7.227	-0.2419910d+22
7.062	-0.1472302d+22
6.897	-0.4954962d+21
6.733	0.5109825d+21
6.568	0.1547626d+22
6.403	0.2614948d+22
6.238	0.3713480d+22
6.073	0.4843779d+22
5.909	0.6006427d+22
5.744	0.7202031d+22
5.579	0.8431229d+22
5.414	0.9694689d+22
5.249	0.1099311d+23
5.085	0.1232725d+23
4.920	0.1369788d+23
4.755	0.1510583d+23
4.590	0.1655199d+23
4.426	0.1803731d+23
4.261	0.1956279d+23
4.096	0.2112952d+23
3.931	0.2273868d+23
3.766	0.2439155d+23
3.602	0.2608953d+23
3.437	0.2783414d+23
3.272	0.2962710d+23
3.107	0.3147031d+23
2.942	0.3336592d+23
2.778	0.3531641d+23
2.613	0.3732462d+23
2.448	0.3939391d+23
2.283	0.4152830d+23
2.118	0.4373272d+23
1.954	0.4601335d+23
1.789	0.4837818d+23
1.624	0.5083798d+23
1.459	0.5340791d+23
1.294	0.5611061d+23
1.130	0.5898259d+23
0.965	0.6208896d+23
0.800	0.6556339d+23

BETA=   0.65643740334535 RAD./MM
NORMALIZED BETA=   2.61076
NORMALIZED WAVE-LENGTH=   0.38303
E EFFECTIVE=   6.81609

CHECK THE DETERMINANT FOR ANY SPURIOUS MODE.

APPENDIX III COMPUTER PROGRAMS 226

FREQ= 0.1700000d+11   HZ

FICTITIOUS DIELECTRIC CONSTANT	DETERMINANT
8.875	-0.8663854d+22
8.710	-0.7988208d+22
8.545	-0.7281891d+22
8.381	-0.6543959d+22
8.216	-0.5773407d+22
8.051	-0.4969167d+22
7.886	-0.4130099d+22
7.721	-0.3254984d+22
7.557	-0.2342518d+22
7.392	-0.1391301d+22
7.227	-0.3998269d+21
7.062	0.6335270d+21
6.897	0.1710516d+22
6.733	0.2833042d+22
6.568	0.4003172d+22
6.403	0.5223162d+22
6.238	0.6495481d+22
6.073	0.7822837d+22
5.909	0.9208221d+22
5.744	0.1065494d+23
5.579	0.1216668d+23
5.414	0.1374755d+23
5.249	0.1540219d+23
5.085	0.1713582d+23
4.920	0.1895440d+23
4.755	0.2086475d+23
4.590	0.2287475d+23
4.426	0.2499359d+23
4.261	0.2723205d+23
4.096	0.2960297d+23
3.931	0.3212175d+23
3.766	0.3480716d+23
3.602	0.3768239d+23
3.437	0.4077656d+23
3.272	0.4412692d+23
3.107	0.4778219d+23
2.942	0.5180769d+23
2.778	0.5629353d+23
2.613	0.6136851d+23
2.448	0.6722456d+23
2.283	0.7416295d+23
2.118	0.8268884d+23
1.954	0.9372608d+23
1.789	0.1091808d+24
1.624	0.1337633d+24
1.459	0.1833138d+24
1.294	0.3668190d+24
1.130	-0.5562912d+24
0.965	-0.7424640d+23

APPENDIX III COMPUTER PROGRAMS

```
                       0.800                -0.5770314d+20

BETA=     0.95337325816051  RAD./MM
NORMALIZED BETA=     2.67651
NORMALIZED WAVE-LENGTH=    0.37362
E EFFECTIVE=    7.16373


BETA=     0.38740810628418  RAD./MM
NORMALIZED BETA=     1.08762
NORMALIZED WAVE-LENGTH=    0.91944
E EFFECTIVE=    1.18291

CHECK THE DETERMINANT FOR ANY SPURIOUS MODE.
```

APPENDIX III COMPUTER PROGRAMS   228

III.7 Listing of MGNTW

```
C*************************************************************
C                                                             *
C   THIS SUBROUTINE COMPUTES THE SPECTRAL DOMAIN GREEN'S      *
C   FUNCTION OF PLANAR TRANSMISSION LINES OF GENERAL          *
C   CROSS-SECTION SHOWN IN FIG.III.2.a OF THIS APPENDIX.      *
C   FOR FURTHER DETAILS, SEE SECTIONS 6.2.3 AND III.2.        *
C                                                             *
      SUBROUTINE MGNTW(BET,FREQ,GG11,GG12,GG21,GG22)
C                                                             *
C*************************************************************
      IMPLICIT REAL*8(A-H,O-Z)
      DIMENSION GG11(200),GG12(200),GG21(200),GG22(200)
      COMMON/XXX/E0,ER1,ER2,AM0,H1,TH,D,A,NN
      COMMON PI
      ER02=ER2*E0
      OMEGA=2.*PI*FREQ
      AK0=OMEGA*DSQRT(AM0*E0)
      AK1=AK0*DSQRT(ER1)
      AK2=AK0*DSQRT(ER2)
      SK0=AK0**2
      SK1=AK1**2
      SK2=AK2**2
      SBET=BET**2
      AKB0=SK0-SBET
      AKB1=SK1-SBET
      AKB2=SK2-SBET
      AKOB=AKB0/BET
      SKK=SK1-SK2
      DO 100 N=1,NN
      AN=N-1
      ALPN=AN*PI/A
      SALPN=ALPN**2
C     ---------------------
      ABA=SALPN+SBET-SK2
      GAMA2=DSQRT(DABS(ABA))
      GH2=GAMA2*TH
      IF(ABA) 50,50,51
   50 TA2=DSIN(GH2)/DCOS(GH2)
      GO TO 52
   51 TA2=DTANH(GH2)
   52 ACA=SALPN+SBET-SK1
      ADA=SALPN+SBET-SK0
      GAMA1=DSQRT(DABS(ACA))
      GAMA0=DSQRT(DABS(ADA))
      GH1=GAMA1*H1
      GD=GAMA0*D
C     ---------------------
      GAA=GAMA1/GAMA2
      EG12=GAA*ER1/ER2
      ABOG=ALPN*BET/(OMEGA*GAMA2)
```

APPENDIX III COMPUTER PROGRAMS

```
          ABOGE=ABOG/ER02
          ABOGM=ABOG/AM0
          OBG2=OMEGA*BET*GAMA2
          OBG2E=OBG2*ER02
          OBG2M=OBG2*AM0
          OBG0E=OMEGA*E0*GAMA0/BET
          OBG0M=OMEGA*AM0*GAMA0/BET
          AKG2=SK2*GAMA2**2
          AMUL=SK2*AKB1-SALPN*SK1
C         ------------------------
          IF(ACA) 60,60,61
   61     TA1=DTANH(GH1)
          A11=(AKB1+EG12*AKB2*TA1*TA2)/BET
          AL1=-ALPN*(1.D0+EG12*TA2*TA1)
          G11=(AMUL*TA2-AKG2*EG12*TA1)/OBG2M
          GO TO 62
   60     TA1=DSIN(GH1)/DCOS(GH1)
          A11=(AKB1-EG12*AKB2*TA1*TA2)/BET
          AL1=-ALPN*(1.D0-EG12*TA1*TA2)
          G11=(AMUL*TA2+AKG2*EG12*TA1)/OBG2M
   62     B11=ABOGE*SKK*TA2*TA1/BET
          C11=-ABOGM*SKK*TA2/BET
          D11=(AKB1*TA1+GAA*AKB2*TA2)/BET
          F11=-(AMUL*TA2*TA1-AKG2*GAA)/OBG2E
          H11=-ALPN*(TA1+GAA*TA2)
C         -----------------------------------
          IF(ADA) 63,63,64
   64     CTAD=1.D0/DTANH(GD)
          H0=-ALPN*CTAD
          B0=AKOB*CTAD
          GO TO 65
   63     CTAD=DCOS(GD)/DSIN(GD)
          H0=ALPN*CTAD
          B0=-AKOB*CTAD
   65     A0=AKOB
          C0=-ALPN
          D0=-OBG0M
          AL0=OBG0E*CTAD
C         --------------------
          ADBC=A11*D11+B11*C11
          AHBG=A11*H11+B11*G11
          BLAF=B11*AL1-A11*F11
          CHGD=C11*H11-G11*D11
          CFLD=C11*F11+AL1*D11
          GFLH=G11*F11+AL1*H11
C         --------------------
          CH0=C0*H0
          DL0=D0*AL0
          CB0=C0*B0
          BL0=AL0*B0
          AH0=A0*H0
          AD0=A0*D0
```

APPENDIX III COMPUTER PROGRAMS

```
          AB0=A0*B0
C         --------------------
          DET=ADBC*(CH0+DL0)
        *  -AHBG*CB0+BLAF*BL0+CHGD*AD0
        *  -CFLD*AH0+GFLH*AB0
          AM1=(AHBG*D0+BLAF*H0)/DET
          AN1=-(BLAF*B0+ADBC*D0)/DET
          AM2=(GFLH*A0+BLAF*AL0-AHBG*C0)/DET
          AN2=(ADBC*C0-CFLD*A0)/DET
          DD11=-AKOB*AM1
          DD12=AKOB*AN1
          DD21=ALPN*AM1+OBG0M*AM2
          DD22=-ALPN*AN1-OBG0M*AN2
          DETG=DD11*DD22-DD12*DD21
          GG11(N)=DD22/DETG
          GG12(N)=-DD12/DETG
          GG21(N)=-DD21/DETG
          GG22(N)=DD11/DETG
          IF(N-1) 100,79,100
       79 GG12(N)=GG12(N)/2.D0
C         --------------------
      100 CONTINUE
          RETURN
          END
```

APPENDIX III COMPUTER PROGRAMS

III.8 Listing of ANPDL

```
C***********************************************************
C                                                           *
C   THIS SUBROUTINE COMPUTES THE SPECTRAL DOMAIN GREEN'S    *
C   FUNCTION OF PLANAR TRANSMISSION LINES OF GENERAL        *
C   CROSS-SECTION SHOWN IN FIG.III.2.b OF THIS APPENDIX.    *
C   FOR FURTHER DETAILS, SEE SECTIONS 6.2.3 AND III.3.      *
C                                                           *
      SUBROUTINE ANPDL(BET,FREQ,GG11,GG12,GG21,GG22)
C                                                           *
C***********************************************************
      IMPLICIT REAL*8(A-H,O-Z)
      DIMENSION GG11(200),GG12(200),GG21(200),GG22(200)
      COMMON/XXX/E0,ER1,ER2,AM0,H1,TH,D,A,NN
      COMMON PI
      ER02=ER2*E0
      OMEGA=2.D0*PI*FREQ
      AK0=OMEGA*DSQRT(AM0*E0)
      AK1=AK0*DSQRT(ER1)
      AK2=AK0*DSQRT(ER2)
      SK0=AK0**2
      SK1=AK1**2
      SK2=AK2**2
      SBET=BET**2
      AKB0=SK0-SBET
      AKB1=SK1-SBET
      AKB2=SK2-SBET
      AKOB=AKB0/BET
      SKK=SK1-SK2
      DO 100 N=1,NN
      AN=N-1
      ALPN=AN*PI/(2.D0*A)
      SALPN=ALPN**2
C     ----------------------
      ABA=SALPN+SBET-SK2
      GAMA2=DSQRT(DABS(ABA))
      GH2=GAMA2*TH
      IF(ABA) 50,50,51
   50 TA2=DSIN(GH2)/DCOS(GH2)
      GO TO 52
   51 TA2=DTANH(GH2)
   52 ACA=SALPN+SBET-SK1
      ADA=SALPN+SBET-SK0
      GAMA1=DSQRT(DABS(ACA))
      GAMA0=DSQRT(DABS(ADA))
      GH1=GAMA1*H1
      GD=GAMA0*D
C     ---------------
      GAA=GAMA1/GAMA2
      EG12=GAA*ER1/ER2
      ABOG=ALPN*BET/(OMEGA*GAMA2)
```

APPENDIX III COMPUTER PROGRAMS

```
            ABOGE=ABOG/ER02
            ABOGM=ABOG/AM0
            OBG2=OMEGA*BET*GAMA2
            OBG2E=OBG2*ER02
            OBG2M=OBG2*AM0
            OBG0E=OMEGA*E0*GAMA0/BET
            OBG0M=OMEGA*AM0*GAMA0/BET
            AKG2=SK2*GAMA2**2
            AMUL=SK2*AKB1-SALPN*SK1
            IF(N-2*(N/2).EQ.0) GO TO 40
C           ---------------------------------
            IF(ACA) 60,60,61
     61     TA1=DTANH(GH1)
            A11=(AKB1+EG12*AKB2*TA1*TA2)/BET
            AL1=-ALPN*(1.D0+EG12*TA2*TA1)
            G11=(AMUL*TA2-AKG2*EG12*TA1)/OBG2M
            GO TO 62
     60     TA1=DSIN(GH1)/DCOS(GH1)
            A11=(AKB1-EG12*AKB2*TA1*TA2)/BET
            AL1=-ALPN*(1.D0-EG12*TA1*TA2)
            G11=(AMUL*TA2+AKG2*EG12*TA1)/OBG2M
     62     B11=ABOGE*SKK*TA2*TA1/BET
            C11=-ABOGM*SKK*TA2/BET
            D11=(AKB1*TA1+GAA*AKB2*TA2)/BET
            F11=-(AMUL*TA2*TA1-AKG2*GAA)/OBG2E
            H11=-ALPN*(TA1+GAA*TA2)
            GO TO 42
C           ---------------------------------
     40     IF(ACA) 70,70,71
     71     TA1=DTANH(GH1)
            D11=(AKB1+GAA*AKB2*TA1*TA2)/BET
            F11=-(AMUL*TA2-AKG2*GAA*TA1)/OBG2E
            H11=-ALPN*(1.D0+GAA*TA2*TA1)
            GO TO 72
     70     TA1=DSIN(GH1)/DCOS(GH1)
            D11=(AKB1-GAA*AKB2*TA1*TA2)/BET
            F11=-(AMUL*TA2+AKG2*GAA*TA1)/OBG2E
            H11=-ALPN*(1.D0-GAA*TA2*TA1)
     72     A11=(AKB1*TA1+EG12*AKB2*TA2)/BET
            B11=ABOGE*SKK*TA2/BET
            C11=-ABOGM*SKK*TA2*TA1/BET
            AL1=-ALPN*(TA1+EG12*TA2)
            G11=(AMUL*TA2*TA1-AKG2*EG12)/OBG2M
C           ---------------------------------
     42     IF(ADA) 63,63,64
     64     CTAD=1.D0/DTANH(GD)
            H0=-ALPN*CTAD
            B0=AKOB*CTAD
            GO TO 65
     63     CTAD=DCOS(GD)/DSIN(GD)
            H0=ALPN*CTAD
            B0=-AKOB*CTAD
```

APPENDIX III COMPUTER PROGRAMS

```
   65 A0=AKOB
      C0=-ALPN
      D0=-OBG0M
      AL0=OBG0E*CTAD
C     -------------------
      ADBC=A11*D11+B11*C11
      AHBG=A11*H11+B11*G11
      BLAF=B11*AL1-A11*F11
      CHGD=C11*H11-G11*D11
      CFLD=C11*F11+AL1*D11
      GFLH=G11*F11+AL1*H11
C     -------------------
      CH0=C0*H0
      DL0=D0*AL0
      CB0=C0*B0
      BL0=AL0*B0
      AH0=A0*H0
      AD0=A0*D0
      AB0=A0*B0
C     ------------------------------------
      DET=ADBC*(CH0+DL0)-AHBG*CB0+BLAF*BL0
     *    +CHGD*AD0-CFLD*AH0+GFLH*AB0
      AM1=(AHBG*D0+BLAF*H0)/DET
      AN1=-(BLAF*B0+ADBC*D0)/DET
      AM2=(GFLH*A0+BLAF*AL0-AHBG*C0)/DET
      AN2=(ADBC*C0-CFLD*A0)/DET
      DD11=-AKOB*AM1
      DD12=AKOB*AN1
      DD21=ALPN*AM1+OBG0M*AM2
      DD22=-ALPN*AN1-OBG0M*AN2
      DETG=DD11*DD22-DD12*DD21
      GG11(N)=DD22/DETG
      GG12(N)=-DD12/DETG
      GG21(N)=-DD21/DETG
      GG22(N)=DD11/DETG
      IF(N-1) 100,79,100
   79 GG12(N)=GG12(N)/2.D0
C     -------------------
  100 CONTINUE
      RETURN
      END
```

APPENDIX III COMPUTER PROGRAMS

III.9 Listing of MSCL2

```
C*************************************************************
C                                                             *
C    THIS SUBROUTINE COMPUTES THE SPECTRAL DOMAIN GREEN'S     *
C    FUNCTION OF PLANAR TRANSMISSION LINES OF GENERAL CROSS-  *
C    SECTION SHOWN IN FIG.III.2.c OF THIS APPENDIX.           *
C    FOR FURTHER DETAILS, SEE SECTIONS 6.3.1 AND III.4.       *
C                                                             *
      SUBROUTINE MSCL2(BET,FREQ,GG11,GG12,GG13,GG14,
     *                     GG22,GG23,GG31,GG32,GG41)
C*************************************************************
      IMPLICIT REAL*8(A-H,O-Z)
      DIMENSION GG11(200),GG12(200),GG13(200),GG14(200)
      DIMENSION GG22(200),GG23(200),GG31(200),GG32(200),GG41(200)
      COMMON/XXX/E0,AM0,ER1,ER2,ER3,TH,D,H,A,NN,ICODE
      COMMON PI
      ER01=ER1*E0
      ER02=ER2*E0
      ER03=ER3*E0
      OMEGA=2.D0*PI*FREQ
      OA=OMEGA*AM0
      OE1=OMEGA*ER01
      OE2=OMEGA*ER02
      OE3=OMEGA*ER03
C     --------------
      SK1=OA*OE1
      SK2=OA*OE2
      SK3=OA*OE3
      SBET=BET**2
      PBK1=SBET*SK1
      PBK2=SBET*SK2
      SKB1=SK1-SBET
      SKB2=SK2-SBET
      SKB3=SK3-SBET
      SK21=SKB2-SKB1
      SKK=SK1*SK2
      SSK=SK21**2
      BK1=PBK1-SKK
      BK2=PBK2-SKK
C     -------------
      DO 1 N=1,NN
      IF(ICODE.EQ.1) GO TO 8
      AN=N-1
      ALPN=AN*PI/A
      GO TO 9
    8 AN=N
      ALPN=(AN-0.5D0)*PI/A
    9 SALPN=ALPN**2
      AB=ALPN*BET
      AB2=AB*BET
      AB3=AB2*BET
```

APPENDIX III COMPUTER PROGRAMS

```
      SAB=AB**2
      SKA1=SK1-SALPN
      SKA2=SK2-SALPN
      SKA3=SK3-SALPN
      PAK1=SALPN*SK1
      PAK2=SALPN*SK2
C     ----------------
      SGAMA1=SALPN-SKB1
      SGAMA2=SALPN-SKB2
      SGAMA3=SALPN-SKB3
      GAMA1=DSQRT(DABS(SGAMA1))
      GAMA2=DSQRT(DABS(SGAMA2))
      GAMA3=DSQRT(DABS(SGAMA3))
C     --------------------------
      SG1=SGAMA1/GAMA1
      SSG1=SG1/GAMA1
      SG3=SGAMA3/GAMA3
C     ----------------
      OMA1=SG1*OA
      OMA2=GAMA2*OA
      OME2=GAMA2*OE2
      OMA3=SG3*OA
C     --------------
      GK2=GAMA2*SK2
      OTC=GK2*SG1
      OTC2=GK2*OA
      OT2C=OTC2*SG1*GAMA2
      SOTC=OTC**2
      SABK=SAB*SK21
      X1=BK1+SALPN*SK2
      X2=BK2+SALPN*SK1
      ZA=SKK*SGAMA1*SGAMA2
      ZB=SKB2*X1+SABK
      ZC=SKA2*X2+SABK
      ZD=SK1*SGAMA1+SSK
      ZE=SK2*SGAMA1+SK1*SGAMA2
C     ------------------------
      GH=GAMA1*H
      GT=GAMA2*TH
      GD=GAMA3*D
C     ----------------------
      IF(SGAMA1.LE.0) GO TO 2
      TA1=DTANH(GH)
      GO TO 3
    2 TA1=DTAN(GH)
    3 IF(SGAMA2.LE.0) GO TO 4
      TA2=DTANH(GT)
      TCH2=DSQRT(1.D0-TA2**2)
      GO TO 5
    4 TA2=DTAN(GT)
      COS2=DCOS(GT)
      RC2=COS2/DABS(COS2)
```

APPENDIX III COMPUTER PROGRAMS

```
          TCH2=DSQRT(1.D0+TA2**2)/RC2
     5    IF(SGAMA3.LE.0) GO TO 6
          TA3=DTANH(GD)
          GO TO 7
     6    TA3=DTAN(GD)
C         ------------
     7    TC=TA2/TA1
          T2C=TC*TA2
          TC2=TC/TA1
          STC=TC**2
C         ------------
          XT2=TA2/OMA2
          XC1=1.D0/(OMA1*TA1)
          XTC2=TC2/(OTC2*SSG1)
          XT2C=T2C/OT2C
          XTC=TC/OTC
          XSTC=STC/SOTC
          TE2=TA2/OME2
          CA3=1.D0/(OMA3*TA3)
C         ----------------------
          EDFC=-AB2*(XT2+XC1+SK1*XTC2+ZD*XT2C)
          FAEB=-BET*(1.D0+ZE*XTC+ZA*XSTC)
          BCAD=SBET*(SKB2*XT2+SKB1*XC1+SK1*SKB2*XTC2+ZB*XT2C)
          EHFG=SKA2*XT2+SKA1*XC1+SK1*SKA2*XTC2+ZC*XT2C
C         ---------------------------------------------
          DET=-BET*FAEB
C         ---------------
          GG11(N)=SBET*(1.D0+X2*XTC)*TCH2/DET
          GG12(N)=AB3*SK21*XTC*TCH2/DET
          GG13(N)=AB3*(1.D0+SK2*SGAMA2*XTC)*TE2/DET
          GG14(N)=-SBET*(SKB2+SK2*SKB1*SGAMA2*XTC)*TE2/DET
          GG22(N)=SBET*(1.D0+X1*XTC)*TCH2/DET
          GG23(N)=-SBET*(SKA2+SK2*SKA1*SGAMA2*XTC)*TE2/DET
          GG31(N)=(BET*EDFC+AB2*FAEB*CA3)/DET
          GG32(N)=(BET*FAEB*SKB3*CA3-BCAD)/DET
          GG41(N)=(-SBET*EHFG+BET*FAEB*SKA3*CA3)/DET
C         ------------------------------------------
          IF(ICODE.EQ.1) GO TO 1
          IF(N.GT.1) GO TO 1
          GG11(N)=GG11(N)/2.D0
          GG12(N)=GG12(N)/2.D0
          GG13(N)=GG13(N)/2.D0
          GG14(N)=GG14(N)/2.D0
          GG22(N)=GG22(N)/2.D0
          GG23(N)=GG23(N)/2.D0
          GG31(N)=GG31(N)/2.D0
          GG32(N)=GG32(N)/2.D0
          GG41(N)=GG41(N)/2.D0
C         -------------------
     1    CONTINUE
          RETURN
          END
```

APPENDIX III COMPUTER PROGRAMS

References

1. Mirshekar-Syahkal D. and Davies J.B., "An accurate, unified solution to various finline structures, of phase constants, characteristic impedance and attenuation", *IEEE Trans. Microwave Theory Tech.*, **MTT-30**, pp.1854-1861, 1982.

2. Mirshekar-Syahkal D. and Davies J.B., "Accurate analysis of coupled strip-finline structure for phase constant, characteristic impedance, dielectric and conductor losses", *ibid*, **MTT-30**, pp.906-910, 1982.

3. Mirshekar-Syahkal D. and Jia B., "Analysis of bilateral finline couplers", *Electron. Lett.*, **Vol. 23**, No. 1, pp.577-579, 1987.

4. Hoefer W.J.R., "Accelerated spectral domain analysis of E-plane circuits suitable for computer aided design", *URSI International Symposium on Electromagnetic Theory*, pp.495-497, 1986.

Index

A
Acceleration of computation of matrix elements 70
Algebraic equation 34
Anisotropic dielectric layer 177
Anisotropic material (s) 173, 175
Antenna
 aperture coupled microstrip 158
 microstrip 5, 169
 microstrip patch 150, 158
 planar 150
 rectangular microstrip 152, 154, 160
 rectangular patch 158
 slot 5
Approximate currents and/or fields 60, 126
Approximate solution 60, 126
Approximation
 first-order 63
 quasi-TEM 75, 79, 99, 107
 TEM 75
 zero-order 63
Array of planar radiators 158
Asymptotic expressions 136
Asymptotic solution 16, 54
Auxiliary functions 168

B
Basis functions 39, 41, 53-57, 60, 64-65, 69, 79, 82, 85-86, 106-107, 110-111, 116, 120, 123, 146, 155, 158, 165, 168, 170-171,
 for five coupled microstrip line 84, 86
 Fourier transforms of 206, 209
 junction 168
 non-singular 59
Bends 5
Bessel functions 55

Boundary conditions 11-12, 25-30, 32, 34, 76-78, 84, 93, 97, 115, 118, 120, 144, 166, 177
 in the Fourier damain 29, 93
 in the space domain 29

C
Characteristic equation 38-39, 62, 147, 173
Characteristic impedance 37, 54
Characteristic matrix 120, 126, 147
Closed-form expression 37
Coaxial feed probe 152
Coefficient matrix 47, 64-65, 70
Complete set 43, 55
Complex permittivity 172
Complex resonant frequency 154
Computer aided design 1
Conducting wedge 15, 17
Conformal mapping 104
Constitutive relations 11-12
Convergence 41
 relative 15, 54
Coplanar waveguide 2, 53, 104-105, 110, 178
 coupled 84
 shielded 199
Coupled differential equations 175
Coupled microstrip coplanar lines 125
Coupled microstrip-finline 126
Coupled microstrip line 7, 40, 110
Coupled mode theory 84, 172
Coupled slot finline 109-110
Cut-off frequency 105

D
Dependence of solution on matrix order 68
Dielectric losses 172
Dielectric wedge 15

Index

Dirichlet's conditions 186
Discontinuity 5
　microstrip 7
　problem 171
Discretization
　non-uniform 41
　uniform 41
Dispersion characteristic
　of antipodal finline 116
　of coplanar waveguide 105, 106
　of inverted microstrip line 107, 110
Distributed amplifier 2
Dyadics 12

E

E-plane circuits 70
E-plane rectangular metal stub 168
Edge condition 11, 54-56, 59, 166
Edge singularity 54
Eigenvalue 70
Electric and magnetic potential functions in the Fourier domain 29
Electric wall 103, 105, 114-115, 125
Electromagnetic shielding 23
Energy 15
Expansion functions 56-57, 82

F

Ferrite 23, 37, 89, 172
Fictitious dielectric constant 203
Field components in the Fourier domain 26, 29
Field singularities 15
Filters 160
Finline 2, 8, 53, 71, 108-116
　antipodal 112, 114, 116, 210
　bilateral 57, 112-114, 175, 210
　coupled slot 110
　coupled strip 121, 125
　coupled unilateral 201
　discontinuities 171
　isolator 89
　model for 109-110
　unilateral 110-114
　with periodic stubs 166
Formal solution 60
Fourier
　coefficients 185
　domain 7
　integral 27, 185
　integral expansion 189
　parameter 26
　series 26, 84, 162-163, 185
　series of odd and even functions 186

Fourier transform 28, 30, 34, 39, 56, 76, 78-79, 83, 93, 95, 98, 143, 145, 152, 164, 189
　closed-form 55, 83
　coefficient 27
　finite 27, 175, 188-189
　inverse 148, 188-189
　two-dimensional finite 145, 163
Fourier transformation 57
Fourier transforms
　of boundary conditions 97
　of derivatives of a function 189
　of potential functions 94
Frequency selective surfaces 168
Full-wave
　analysis 81
　assumption 91
　solution 25, 75, 91
　spectral domain analysis 79, 81

G

Galerkin method (technique) 41, 45-49, 65, 79, 84, 120, 126, 146-147, 155, 165
Gauss elimination technique 209
Green's function 36, 75, 92, 99, 105, 110, 112, 123, 177, 191, 194, 196
　antipodal finline 114
　dyadic 37
　full-wave spectral domain 129
　matrix 64, 91
　of antipodal finline 114, 116
　of bilateral finline 115
　of coplanar waveguide 91
　of coupled microstrip finline 91
　of finline 91
　of multilayer structures 91-92, 95, 99
　of unilateral finline 115
　spectral domain 36, 40, 53, 66, 68-69, 71, 81, 99, 107, 120, 122-123, 134
　unilateral finline 114

H

Half range expansion 187
Hankel transform 143
Helmholtz equation 19-20

I

Immittance method (approach) 7, 91, 129
Impedance
　characteristic 37, 54
　driving point 133, 136
　input 5, 152, 157-158
　transfer 136

Index

Incident field 154, 169-170
Inner product 43, 45-46, 79, 155, 157
Integral equation 31, 34, 36-40, 57, 81
   convolution 36
   eigenvalue 192
   Fredholm's 36, 78
   in the Fourier domain 57
   linear 191
   of the first kind (Fredholm's) 191, 195-196
   of the first kind (Volterra) 192
   of the second kind (Fredholm's) 191, 192
   of the second kind (Volterra) 192
Integral transformation 27, 29
Interconnecting stripline 84
Isolators 172
Iteration process 71
Iterative solution 70

J
Junction basis function 168
Junctions 5

K
Kernel 36
   degenerate (separable) 192-193
   function 86
   function of x-s 193
   Hermitian 193
   nonseparable 193
   positive definite 193
   symmetric 193

L
Laplace transform
   two-sided (double sided) 148, 152
Laplace's equation 76
Least-squares technique 171
Legendre polynomials 40, 59-61, 68, 106, 125, 207
   Fourier transforms of 60
   recurrence relation for 60
Linear
   equations 38
   integral equation 191
   operator 43

M
Magnetic wall 99, 104-105, 112, 115, 125
Matrix
   condition 44, 54
   inversion 157
   transfer (chain) 95
Maxwell's equations 11, 93

Microstrip
   models 24
   structure with sinusoidally varying strip 165
Microstrip couplers 89
Microstrip line 2, 5, 7, 13, 23, 29, 31, 37-40, 49, 62-63, 65, 68, 71, 75-78, 81-82, 84, 91, 99, 107, 110, 112, 126, 143, 160, 163-164, 173, 201
   asymmetric coupled 81, 84
   balanced 99
   coupled 7, 40, 110
   even/odd mode in five coupled 84
   inverted 107, 110
   laterally open 24
   multiconductor 84
   open 24, 40, 60
   shielded 24
   transverse equivalent transmission line model of 132
   with single or coupled strips (shielded) 199
   with sinusoidally varying strip 162
   with tuning septums 80
Modal analysis 171
Modal expansion 171
Mode
   discrete 24, 39
   dominant 60
   even 26-27, 40, 56, 59, 65, 85, 105, 112, 125, 145, 202
   Floquet's 162
   higher order 40, 60
   hybrid 20, 25, 91
   LSE 22, 129-130, 132-133
   LSM 22, 129-130, 132-133
   microstrip 125-126
   odd 27, 56, 65, 105, 114, 125, 202
   slot 125-126
   spectral LSE 130, 133, 135
   spectral LSM 130, 133, 136
   surface 136
   TE 19-20, 22, 25, 91, 111, 129-130
   TM 19-20, 22, 25, 91, 111, 129-130
Modes of isolated conducting strip 60
Moment method 41, 43, 45, 54, 120
Monolithic microwave integrated circuits (MMICs) 1
   distributed amplifier 4
Muller's algorithm 173
Multilayer
   multiconductor structures 89-90, 119
   structure 114
   transmission lines 117, 119

Index

Multistrip
   planar structures 86
   single substrate transmission lines 75, 81
   structure 75, 81

N
Nontrivial solution 39, 47, 126, 153, 165

P
Parseval's identity 41, 46, 48, 79, 155, 190
Periodic function 185-186
Periodic structure 8, 160, 163-164
   microstrip 160
   planar 160, 166
Permeability tensor 178
Planar
   phase shifters 172
   radiators 8
   slow wave structures 172
   tapers 171
Planar transmission line 199
   expressions for the spectral domain Green's function of three layer 100
   multilayer multiconductor 7, 141
   nonuniform 171
   periodic 161
   three-layer 100, 199
Point matching method 38-41, 43, 45, 57
Potential
   electric vector 19
   Hertzian vector 19
   magnetic vector 19
   scalar 144
Potential function (s) 19, 21, 27, 76-77, 91, 93, 114-115
   coefficients 156
   electric 130
   Fourier domain 175
   in the Fourier domain 26, 28
   magnetic 130
   scalar 25, 91, 160, 178
   scalar electric 20-21
   scalar magnetic 21
   spectral domain 77
Pulse functions 41, 57

Q
Quasi-static approximation 104
Quasi-TEM 63, 71
   approximation 75, 79, 99, 107
   assumption 80, 91
   solution 75, 104
   spectral domain formulation 75

R
Radiation 5, 150
Reciprocity theorem 155-156
Regular field/current
   components 55, 59
   distributions 207
Relative convergence 15, 54
Resonator
   disk 141, 143
   isosceles triangular patch 149
   microstrip line 141, 143
   microstrip patch 150
   open 149
   open rectangular planar 152
   planar 8, 141, 152
   rectangular patch 152
   ring 141, 143
   triangular patch 141, 149

S
Scattering 168
   parameters 171
   problems 168
Set of pulses 57
Sharp edge 14, 41, 53-54
Singular field/current
   components 55, 59
   distributions 207
Singular function (s) 56
Singular points 148
Slot fields 127
Slot line 110
   coupled 110
   with single or coupled slots (shielded) 199
Slow wave devices 160
Solution
   non-physical 54
   nontrivial 39
   spurious 55, 206
Space domain 30, 34
Space harmonics 163
Spectral domain
   full-wave analysis 104
   immittance approach 91, 129, 134
   potential functions 114-115
Spectral domain Green's function
   matrix 66
   of antipodal finline 114-116, 199
   of bilateral finline 113-114, 199
   of microstrip resonator 146
   of multilayer multiconductor structure 134
   of multilayer multiconductor transmission line 91

Index

of multilayer planar transmission line 136
of shielded coupled strip-slot line 199
of three-layer planar transmission line with magnetic wall 210
of three-layer planar transmission line with rotational symmetry 210
of three-layer structure 207
Spectral parameter 77, 103, 105, 145, 169
Spectrum
 continuous 48
 discrete 48
Spurious solution 55
Strip current 127
Strip edge 54, 166
Striplines with periodic stubs 166
Substrate
 anisotropic 24, 172
 ferrite 178
 lossy 24, 172
 semi-conductor 8, 23, 172
 semi-insulating 8
Surface charge density 77
Surface current density 30
Surface wave 149
 poles 154
Suspended microstrip line with tuning septums 127
Symmetry
 electric/magnetic wall 84
 rotational 114-115

T

TEM (quasi-) 75
 approximation 76
 solution 71
Testing functions 44-45
Three-layer structure 115
Transfer matrix 91-92, 95
 approach (method) 7, 91, 99, 117, 129
 equation 93
Transverse equivalent transmission line model 129, 134-135
Trial functions 43, 121
Trigonometric functions 40, 56, 59, 68, 123, 185, 207
Trivial solution 39

U

Unilateral finlines with single or coupled gaps 199

V

Vector expansion 166

W

Wave
 admittances 131
 equation 19, 21, 25, 28-29, 144, 164
Waveguide
 cylindrical 19-20
 double ridge 111
 inhomogeneously filled 22
 ridge 108, 111

Z

Zero-order
 approximation 63
 solution 68